Lecture Notes in Computer Science 13222

More information about this series at https://link.springer.com/bookseries/558

Leslie Pérez Cáceres · Sébastien Verel (Eds.)

Evolutionary Computation in Combinatorial Optimization

22nd European Conference, EvoCOP 2022
Held as Part of EvoStar 2022
Madrid, Spain, April 20–22, 2022
Proceedings

 Springer

Editors
Leslie Pérez Cáceres ⓘD
Pontificia Universidad Católica de Valparaíso
Valparaíso, Chile

Sébastien Verel ⓘD
Université du Littoral Côte d'Opale
Calais, France

ISSN 0302-9743   ISSN 1611-3349 (electronic)
Lecture Notes in Computer Science
ISBN 978-3-031-04147-1   ISBN 978-3-031-04148-8 (eBook)
https://doi.org/10.1007/978-3-031-04148-8

This Springer imprint is published by the registered company Springer Nature Switzerland AG
The registered company address is: Gewerbestrasse 11, 6330 Cham, Switzerland

Preface

Combinatorial optimization problems are inherently embedded in decision-making tasks, and in such a way they emerge in the most diverse contexts, from day-to-day, to the most complex tasks in scientific, business, and industrial fields. Combinatorial optimization problems require finding a solution, a value assignment for a set of discrete decision variables, which optimizes one or more criteria. Thanks, in part, to the development of techniques that can solve combinatorial optimization problems efficiently, computer-aided decision-making is nowadays common in a wide range of areas such as transportation, scheduling, distribution, planning, etc. The research and development of effective techniques are crucial for tackling new challenges that arise as more complex and massive problems are constantly formulated due to the ever-expanding use of technology.

Evolutionary algorithms and related methods are often inspired by the dynamics observed in biological and other natural processes. These techniques are among the most popular and effective alternatives available to solve large and complex combinatorial optimization problems due to their intuitive, general-purpose, and adaptable nature. The field's success has been possible thanks to the work of dedicated researchers, both in academia and industry, who have developed and constantly improved the techniques over the past decades. New challenges in combinatorial optimization are constantly arising. At the same time, the ever-increasing computational power allows the scientific community to extend the field into new application domains, consider more ambitious goals and develop more sophisticated techniques. The papers in this volume present recent theoretical and experimental advances in combinatorial optimization, evolutionary algorithms, and related research fields.

This volume contains the proceedings of EvoCOP 2022, the 22nd European Conference on Evolutionary Computation in Combinatorial Optimisation. The conference was held in Madrid, Spain, 20–22 April, 2022. The EvoCOP conference series started in 2001, with the first workshop specifically devoted to evolutionary computation in combinatorial optimization. It became an annual conference in 2004. EvoCOP 2022 was organized together with EuroGP (the 25th European Conference on Genetic Programming), EvoMUSART (the 11th International Conference on Computational Intelligence in Music, Sound, Art and Design), and EvoApplications (the 25th European Conference on the Applications of Evolutionary Computation, formerly known as EvoWorkshops), in a joint event collectively known as EvoStar 2022. Previous EvoCOP proceedings were published by Springer in the Lecture Notes in Computer Science series (LNCS volumes 2037, 2279, 2611, 3004, 3448, 3906, 4446, 4972, 5482, 6022, 6622, 7245, 7832, 8600, 9026, 9595, 10197, 10782, 11452, 12102, and 12692). The table on the next page reports the statistics for each of the previous conferences.

This year, 13 out of 28 papers were accepted after a rigorous double-blind process, resulting in a 46% acceptance rate. We would like to acknowledge the quality and timeliness of our high-quality and diverse Program Committee members' work. Each year the members give freely of their time and expertise, in order to maintain high standards in EvoCOP and provide constructive feedback to help authors improve their papers. Decisions considered both the reviewers' report, and the evaluation of the program chairs. The 13 accepted papers cover a variety of topics, ranging from the foundations of evolutionary computation algorithms and other search heuristics, to their accurate design and application to combinatorial optimization problems. Fundamental and methodological aspects deal with runtime analysis, the study of metaheuristics core components, the clever design of their search principles, and their careful selection and configuration by means of automatic algorithm configuration and hyper-heuristics. Applications cover problem domains such as scheduling, routing, permutation, and general graph problems. We believe that the range of topics covered in this volume of EvoCOP proceedings reflects the current state of research in the fields of evolutionary computation and combinatorial optimization.

EvoCOP	LNCS vol.	Submitted	Accepted	Acceptance (%)
2022	13222	28	13	46.4
2021	12692	42	14	33.3
2020	12102	37	14	37.8
2019	11452	37	14	37.8
2018	10782	37	12	32.4
2017	10197	39	16	41.0
2016	9595	44	17	38.6
2015	9026	46	19	41.3
2014	8600	42	20	47.6
2013	7832	50	23	46.0
2012	7245	48	22	45.8
2011	6622	42	22	52.4
2010	6022	69	24	34.8
2009	5482	53	21	39.6
2008	4972	69	24	34.8
2007	4446	81	21	25.9
2006	3906	77	24	31.2
2005	3448	66	24	36.4
2004	3004	86	23	26.7
2003	2611	39	19	48.7
2002	2279	32	18	56.3
2001	2037	31	23	74.2

We would like to express our appreciation to the various persons, and institutions making EvoCOP 2022 a successful event. Firstly, we thank the local organization team, led by Iñaki Hidalgo from the Universidad Complutense Madrid, Spain. Our acknowledgments also go to SPECIES, the Society for the Promotion of Evolutionary Computation in Europe and its Surroundings, aiming to promote evolutionary algorithmic thinking within Europe and wider, and more generally to promote inspiration of parallel algorithms derived from natural processes. We extend our acknowledgments to Nuno Lourenço from the University of Coimbra, Portugal, for his dedicated work with the submission and registration system, to João Correia from the University of Coimbra, Portugal, for the EvoStar publicity and social media service, to Francisco Chicano from the University of Málaga, Spain, for managing the EvoStar website, and to Sérgio Rebelo and Tiago Martins from the University of Coimbra, Portugal, for their important graphic design work. We wish to thank our prominent keynote speakers, Gabriela Ochoa from the University of Stirling, UK, and Pedro Larrañaga from the Technical University of Madrid, Spain. Finally, we express our appreciation to Anna I. Esparcia-Alcázar from SPECIES, Europe, whose considerable efforts in managing and coordinating EvoStar helped towards building a unique, vibrant, and friendly atmosphere.

Special thanks also to Christian Blum, Francisco Chicano, Carlos Cotta, Peter Cowling, Jens Gottlieb, Jin-Kao Hao, Jano van Hemert, Bin Hu, Arnaud Liefooghe, Manuel Lopéz-Ibáñez, Peter Merz, Martin Middendorf, Gabriela Ochoa, Luís Paquete, Günther R. Raidl, and Christine Zarges for their hard work and dedication at past editions of EvoCOP, making this one of the reference international events in evolutionary computation and metaheuristics.

March 2022   Leslie Pérez Cáceres
                                                             Sébastien Verel

Organization

Organizing Committee

Conference Chairs

Leslie Pérez-Cáceres	Pontificia Universidad Católica de Valparaíso, Chile
Sébastien Verel	Université du Littoral Côte d'Opale, France

Local Organization

Iñaki Hidalgo	Universidad Complutense Madrid, Spain

Publicity and e-Media Chair

João Correia	University of Coimbra, Portugal

Web Chair

Francisco Chicano	University of Málaga, Spain

Submission and Registration Manager

Nuno Lourenço	University of Coimbra, Portugal

EvoCOP Steering Committee

Christian Blum	Artificial Intelligence Research Institute, Spain
Francisco Chicano	University of Málaga, Spain
Carlos Cotta	University of Málaga, Spain
Peter Cowling	Queen Mary University of London, UK
Jens Gottlieb	SAP AG, Germany
Jin-Kao Hao	University of Angers, France
Jano van Hemert	Optos, UK
Bin Hu	Austrian Institute of Technology, Austria
Arnaud Liefooghe	University of Lille, France
Manuel Lopéz-Ibáñez	University of Málaga, Spain
Peter Merz	Hannover University of Applied Sciences and Arts, Germany

Martin Middendorf	University of Leipzig, Germany
Gabriela Ochoa	University of Stirling, UK
Luís Paquete	University of Coimbra, Portugal
Günther Raidl	Vienna University of Technology, Austria
Christine Zarges	Aberystwyth University, UK

Society for the Promotion of Evolutionary Computation in Europe and Its Surroundings (SPECIES)

Marc Schoenauer (President)
Anna I. Esparcia-Alcázar (Vice-president, Secretary, and EvoStar Coordinator)
Wolfgang Banzhaf (Treasurer)

Program Committee

Maria João Alves	University of Coimbra/INESC Coimbra, Portugal
Khulood Alyahya	University of Birmingham, UK
Matthieu Basseur	Université du Littoral Côte d'Opale, France
Christian Blum	Spanish National Research Council (CSIC), Spain
Alexander Brownlee	University of Stirling, UK
Maxim Buzdalov	ITMO University, Russia
Marco Chiarandini	University of Southern Denmark, Denmark
Francisco Chicano	University of Málaga, Spain
Carlos Coello Coello	CINVESTAV-IPN, Mexico
Nguyen Dang	University of St Andrews, UK
Bilel Derbel	University of Lille, France
Marcos Diez García	University of Exeter, UK
Karl Doerner	University of Vienna, Austria
Benjamin Doerr	Ecole Polytechnique, France
Carola Doerr	Sorbonne University, CNRS, France
Jonathan Fieldsend	University of Exeter, UK
Carlos M. Fonseca	University of Coimbra, Portugal
Carlos García-Martínez	University of Córdoba, Spain
Adrien Goëffon	Universite d'Angers, France
Andreia Guerreiro	University of Coimbra, Portugal
Jin-Kao Hao	University of Angers, France
Mario Inostroza-Ponta	Universidad de Santiago de Chile, Chile
Ekhine Irurozki	Telecom Paris, France
Thomas Jansen	Aberystwyth University, Ireland
Andrzej Jaszkiewicz	Poznan University of Technology, Poland
Marie-Eleonore Kessaci	University of Lille, France
Ahmed Kheiri	Lancaster University, UK

Frederic Lardeux   University of Angers, France
Rhydian Lewis   Cardiff University, UK
Arnaud Liefooghe   University of Lille, France
Jose A. Lozano   University of the Basque Country, Spain
Gabriel Luque   University of Málaga, Spain
Manuel López-Ibáñez   University of Málaga, Spain
Krzysztof Michalak   Wroclaw University of Economics, Poland
Elizabeth Montero   Universidad Andres Bello, Chile
Christine Mumford   University of Cardiff, UK
Nysret Musliu   TU Wien, Austria
Gabriela Ochoa   University of Stirling, UK
Pietro S. Oliveto   University of Sheffield, UK
Beatrice Ombuki-Berman   Brock University, Canada
Luis Paquete   University of Coimbra, Portugal
Mario Pavone   University of Catania, Italy
Paola Pellegrini   IFSTTAR, France
Francisco Pereira   Instituto Superior de Engenharia de Coimbra,
                                         Portugal
Pedro Pinacho   Universidad de Concepción, Chile
Daniel Porumbel   Conservatoire National des Arts et Métiers de
                                         Paris, France
Jakob Puchinger   CentraleSupélec IRT-SystemX, France
Leslie Pérez-Cáceres   Pontificia Universidad Católica de Valparaíso,
                                         Chile
Günther Raidl   Vienna University of Technology, Austria
María Cristina Riff   Universidad Tecnica Federico Santa Maria, Chile
Eduardo Rodriguez-Tello   Cinvestav Tamaulipas, Mexico
Andrea Roli   University of Bologna, Italy
Hana Rudová   Masaryk University, Czech Republic
Valentino Santucci   University for Foreigners of Perugia, Italy
Frédéric Saubion   University of Angers, France
Marcella Scoczynski Ribeiro   Federal University of Technology - Parana, Brazil
  Martins
Patrick Siarry   University of Paris-Est Créteil, France
Jim Smith   University of the West of England, UK
Thomas Stützle   Université Libre de Bruxelles, Belgium
Sara Tari   Université du Littoral Côte d'Opale, France
Renato Tinós   University of São Paulo, Brazil
Nadarajen Veerapen   University of Lille, France
Sébastien Verel   Université du Littoral Côte d'Opale, France
Markus Wagner   University of Adelaide, Australia
Carsten Witt   Technical University of Denmark, Denmark

Takeshi Yamada   NTT, Japan
Christine Zarges   Aberystwyth University, UK
Fangfang Zhang   Victoria University of Wellington, Australia

Contents

On Monte Carlo Tree Search
for Weighted Vertex Coloring

Cyril Grelier, Olivier Goudet, and Jin-Kao Hao[✉]

LERIA, Université d'Angers, 2 Boulevard Lavoisier, 49045 Angers, France
{cyril.grelier,olivier.goudet,jin-kao.hao}@univ-angers.fr

Abstract. This work presents the first study of using the popular Monte Carlo Tree Search (MCTS) method combined with dedicated heuristics for solving the Weighted Vertex Coloring Problem. Starting with the basic MCTS algorithm, we gradually introduce a number of algorithmic variants where MCTS is extended by various simulation strategies including greedy and local search heuristics. We conduct experiments on well-known benchmark instances to assess the value of each studied combination. We also provide empirical evidence to shed light on the advantages and limits of each strategy.

Keywords: Monte Carlo Tree Search · local search · graph coloring · weighted vertex coloring

1 Introduction

The well-known Graph Coloring Problem (GCP) is to color the vertices of a graph using as few colors as possible such that no adjacent vertices share the same color (*legal* or *feasible* solution). The GCP can also be considered as partitioning the vertex set of the graph into a minimum number of color groups such that no vertices in each color group are adjacent. The GCP has numerous practical applications in various domains [14] and has been studied for a long time.

A variant of the GCP called the Weighted Vertex Coloring Problem (WVCP) has recently attracted a lot of interest in the literature [9,16,20,22]. In this problem, each vertex of the graph has a weight and the objective is to find a *legal* solution such that the sum of the weights of the heaviest vertex of each color group is minimized. Formally, given a weighted graph $G = (V, E)$ with vertex set V and edge set E, and let W be the set of weights $w(v)$ associated to each vertex v in V, the WVCP consists in finding a partition of the vertices in V into k color groups $S = \{V_1, \ldots, V_k\}$ $(1 \leq k \leq |V|)$ such that no adjacent vertices belong to the same color group and such that the score $\sum_{i=1}^{k} \max_{v \in V_i} w(v)$ is minimized. One can notice that when all the weights $w(v)$ $(v \in V)$ are equal to one, finding an optimal solution of this problem with a minimum score corresponds to solving the GCP. Therefore, the WVCP can be seen as a more general problem than the GCP and is NP-hard. This extension of the GCP is useful for a number of

L. Pérez Cáceres and S. Verel (Eds.): EvoCOP 2022, LNCS 13222, pp. 1–16, 2022.
https://doi.org/10.1007/978-3-031-04148-8_1

applications in multiple fields such as matrix decomposition problem [18], buffer size management, scheduling of jobs in a multiprocessor environment [17] or partitioning a set of jobs into batches [11].

Different methods have been proposed in the literature to solve the WVCP. First, this problem has been tackled with exact methods: a branch-and-price algorithm [7] and two ILP models proposed in [15] and [5], which transforms the WVCP into a maximum weight independent set problem. These exact methods are able to prove the optimality on small instances, but fail on large instances.

To handle large graphs, several heuristics have been introduced to solve the problem approximately [16,18,20,22]. The first category of heuristics is local search algorithms, which iteratively make transitions from the current solution to a neighbor solution. Three different approaches have been considered to explore the search space: legal, partial legal, or penalty strategies. The legal strategy starts from a *legal* solution and minimizes the score by performing only legal moves so that no color conflict is created in the new solution [18]. The partial legal strategy allows only legal coloring and keeps a set of uncolored vertices to avoid conflicts [16]. The penalty strategy considers both legal and illegal solutions in the search space [20,22], and uses a weighted evaluation function to minimize both the WVCP objective function and the number of conflicts in the illegal solutions. To escape local optima traps, these local search algorithms incorporate different mechanisms such as perturbation strategies [18,20], tabu list [16,20] and constraint reweighting schemes [22].

The second category of existing heuristics for the WVCP relies on the population-based memetic framework that combines local search with crossovers. A recent algorithm [9] of this category uses a deep neural network to learn an invariant by color permutation regression model, useful to select the most promising crossovers at each generation.

Research on combining such learning techniques and heuristics has received increasing attention in the past years [23,24]. In these new frameworks, useful information (e.g., relevant patterns) is learned from past search trajectories and used to guide a local search algorithm.

This study continues on this path and investigates the potential benefits of Monte Carlo Tree Search (MCTS) to improve the results of sequential coloring and local search algorithms. MCTS is a heuristic search algorithm that generated considerable interest due to its spectacular success for the game of Go [8], and in other domains (see a survey of [2] on this topic). It has been recently revisited in combination with modern deep learning techniques for difficult two-player games (cf. AlphaGo [19]). MCTS has also been applied to combinatorial optimization problems seen as a one-player game such as the traveling salesman problem [6] or the knapsack problem [10]. An algorithm based on MCTS has recently been implemented with some success for the GCP in [3]. In this work, we investigate for the first time the MCTS approach for solving the WVCP.

In MCTS, a tree is built incrementally and asymmetrically. For each iteration, a tree policy balancing exploration and exploitation is used to find the most critical node to expand. A simulation is then run from the expanded node and the search tree is updated with the result of this simulation. Its incremental and asymmetric properties make MCTS a promising candidate for the WVCP because in this problem only the heaviest vertex of each color group has an impact on the objective score. Therefore learning to color the heaviest vertices of the graph before coloring the rest of the graph seems particularly relevant for this problem. The contributions of this work are summarized as follows.

First, we present a new MCTS algorithm dedicated to the WVCP, which considers the problem from the perspective of a sequential coloring with a pre-defined vertex order. The exploration of the tree is accelerated with the use of specific pruning rules, which offer the possibility to explore the whole tree in a reasonable amount of time for small instances and to obtain optimality proofs. Secondly, for large instances, when obtaining an exact result is impossible in a reasonable time, we study how this MCTS algorithm can be tightly coupled with other heuristics. Specifically, we investigate the integration of different greedy coloring strategies and a local search procedure with the MCTS algorithm.

The rest of the paper is organized as follows. Section 2 introduces the weighted vertex coloring problem and the constructive approach with a tree. Section 3 describes the MCTS algorithm devised to tackle the problem. Section 4 presents the coupling of MCTS with a local search algorithm. Section 5 reports results of the different versions of MCTS. Section 6 discusses the contributions and presents research perspectives.

2 Constructive Approach with a Tree for the Weighted Graph Coloring Problem

This section presents a tree-based approach for the WVCP, which aims to explore the partial and legal search space of this problem.

2.1 Partial and Legal Search Space

The search space Ω studied in our algorithm concerns legal, but potentially partial, k-colorings. A partial legal k-coloring S is a partition of the set of vertices V into k disjoint independent sets V_i $(1 \leq i \leq k)$, and a set of uncolored vertices $U = V \setminus \bigcup_{i=1}^{k} V_i$. A independent set V_i is a set of mutually non adjacent vertices of the graph: $\forall u, v \in V_i, (u, v) \notin E$. For the WVCP, the number of colors k that can be used is not known in advance. Nevertheless, it is not lower than the chromatic number of the graph $\chi(G)$ and not greater than the number of vertices $|V|$ of the graph. A solution of the WVCP is denoted as partial if $U \neq \emptyset$ and complete otherwise. The objective of the WVCP is to find a complete solution S with a minimum score $f(S)$ given by: $f(S) = \sum_{i=1}^{k} \max_{v \in V_i} w(v)$.

2.2 Tree Search for Weighted Vertex Coloring

Backtracking based tree search is a popular approach for the graph coloring problem [1,12,14]. In our case, a tree search algorithm can be used to explore the partial and legal search space of the WVCP previously defined.

Starting from a solution where no vertex is colored (i.e., $U = V$) and that corresponds to the root node R of the tree, child nodes C are successively selected in the tree, consisting of coloring one new vertex at a time. This process is repeated until a terminal node T is reached (all the vertices are colored). A complete solution (i.e., a legal coloring) corresponds thus to a branch from the root node to a terminal node.

The selection of each child node corresponds to applying a move to the current partial solution being constructed. A move consists of assigning a particular color i to an uncolored vertex $u \in U$, denoted as $< u, U, V_i >$. Applying a move to the current partial solution S, results in a new solution $S \oplus < u, U, V_i >$. This tree search algorithm only considers legal moves to stay in the partial legal space. For a partial solution $S = \{V_1, ..., V_k, U\}$, a move $< u, U, V_i >$ is said legal if no vertex of V_i is adjacent to the vertex u. At each level of the tree, there is at least one possible legal move that applies to a vertex a new color that has never been used before (or putting this vertex in a new empty set V_i, with $k + 1 \leq i \leq |V|$).

Applying a succession of $|V|$ legal moves from the initial solution results in a legal coloring of the WVCP and reaches a terminal node of the tree. During this process, at the level t of the tree ($0 \leq t < |V|$), the current legal and partial solution $S = \{V_1, ..., V_k, U\}$ has already used k colors and t vertices have already received a color. Therefore $|U| = |V| - t$.

At this level, a first naive approach could be to consider all the possible legal moves, corresponding to choosing a vertex in the set U and assigning to the vertex a color i, with $1 \leq i \leq |V|$. This kind of choice can work with small graphs but with large graphs, the number of possible legal moves becomes huge. Indeed, at each level t, the number of possible legal moves can go up to $(|V| - t) \times |V|$. To reduce the set of move possibilities, we consider the vertices of the graph in a predefined order $(v_1, ..., v_{|V|})$. Moreover, to choose a color for the incoming vertex, we consider only the colors already used in the partial solution plus one color (creation of a new independent set). Thus for a current legal and partial solution $S = \{V_1, ..., V_k, U\}$, at most $k + 1$ moves are considered. The set of legal moves is:

$$\mathcal{L}(S) = \{< u, U, V_i >, 1 \leq i \leq k, \forall v \in V_i, (u, v) \notin E\} \cup < u, U, V_{k+1} > \quad (1)$$

This decision cuts the symmetries in the tree while reducing the number of branching factors at each level of the tree.

2.3   Predefined Vertex Order

We propose to consider a predefined ordering of the vertices, sorted by weight then by degree. Vertices with higher weights are placed first. If two vertices have the same weight, then the vertex with the higher degree is placed first. This order is intuitively relevant for the WVCP, because it is more important to place first the vertices with heavy weights which have the most impact on the score as well as the vertices with the highest degree because they are the most constrained decision variables. Such ordering has already been shown to be effective with greedy constructive approaches for the GCP [1] and the WVCP [16].

Moreover, this vertex ordering allows a simple score calculation while building the tree. Indeed, as the vertices are sorted by descending order of their weights, and the score of the WVCP only counts the maximum weight of each color group, with this vertex ordering, the score only increases by the value $w(v)$ when a new color group is created for the vertex v.

3 Monte Carlo Tree Search for Weighted Vertex Coloring

The search tree presented in the last subsection can be huge, in particular for large instances. Therefore, in practice, it is often impossible to perform an exhaustive search of this tree, due to expensive computing time and memory requirements. We turn now to an adaptation of the MCTS algorithm for the WVCP to explore this search tree. MCTS keeps in memory a tree (hereinafter referred to as the MCTS tree) which only corresponds to the already explored nodes of the search tree presented in the last subsection. In the MCTS tree, a leaf is a node whose children have not yet all been explored while a terminal node corresponds to a complete solution. MCTS can guide the search toward the most promising branches of the tree, by balancing exploitation and exploration and continuously learning at each iteration.

3.1 General Framework

The MCTS algorithm for the WVCP is shown in Algorithm 1. The algorithm takes a weighted graph as input and tries to find a legal coloring S with the minimum score $f(S)$. The algorithm starts with an initial solution where the first vertex is placed in the first color group. This is the root node of the MCTS tree. Then, the algorithm repeats a number of iterations until a stopping criterion is met. At every iteration, one legal solution is completely built, which corresponds to walking along a path from the root node to a leaf node of the MCTS tree and performing a simulation (or playout/rollout) until a terminal node of the search tree is reached (when all vertices are colored).

Each iteration of the MCTS algorithm involves the execution of 5 steps to explore the search tree with legal moves (cf. Sect. 2):

1. **Selection.** From the root node of the MCTS tree, successive child nodes are selected until a leaf node is reached. The selection process balances the exploration-exploitation trade-off. The exploitation score is linked to the average score obtained after having selected this child node and is used to guide the algorithm to a part of the tree where the scores are the lowest (the WVCP is a minimization problem). The exploration score is linked to the number of visits to the child node and will incite the algorithm to explore new parts of the tree, which have not yet been explored.
2. **Expansion.** The MCTS tree grows by adding a new child node to the leaf node reached during the selection phase.
3. **Simulation.** From the newly added node, the current partial solution is completed with legal moves, randomly or by using heuristics.

Algorithm 1. MCTS algorithm for the WVCP

1: **Input:** Weighted graph $G = (V, W, E)$
2: **Output:** The best legal coloring S^* found
3: $S^* = \emptyset$ and $f(S^*) = MaxInt$
4: **while** stop condition is not met **do**
5:     $C \leftarrow R$ ▷ Current node corresponding to the root node of the tree
6:     $S \leftarrow \{V_1, U\}$ with $V_1 = \{v_1\}$ and $U = V \backslash V_1$ ▷ Current solution initialized with the first vertex in the first color group
7:     /* Selection */   ▷ Sect. 3.2
8:     **while** C is not a leaf **do**
9:        $C \leftarrow$ select_best_child(C) with legal move $< u, U, V_i >$
10:        $S \leftarrow S \oplus < u, U, V_i >$
11:     **end while**
12:     /* Expansion */   ▷ Sect. 3.3
13:     **if** C has a potential child, not yet open **then**
14:        $C \leftarrow$ open_first_child_not_open(C) with legal move $< u, U, V_i >$
15:        $S \leftarrow S \oplus < u, U, V_i >$
16:     **end if**
17:     /* Simulation */   ▷ Sect. 3.4
18:     complete_partial_solution(S)
19:     /* Update */   ▷ Sect. 3.5
20:     **while** $C \neq R$ **do**
21:        update(C,f(S))
22:        $C \leftarrow$ parent(C)
23:     **end while**
24:     **if** $f(S) < f(S^*)$ **then**
25:        $S^* \leftarrow S$
26:        /* Pruning */   ▷ Sect. 3.6
27:        apply pruning rules
28:     **end if**
29: **end while**
30: return S^*

---

4. **Update.** After the simulation, the average score and the number of visits of each node on the explored branch are updated.
5. **Pruning.** If a new best score is found, some branches of the MCTS tree may be pruned if it is not possible to improve the best current score with it.

The algorithm continues until one of the following conditions is reached:

– there are no more child nodes to expand, meaning the search tree has been fully explored. In this case, the best score found is proven to be optimal.
– a cutoff time is attained. The minimum score found so far is returned. It corresponds to an upper bound of the score for the given instance.

3.2   Selection

The selection starts from the root node of the MCTS tree and selects children nodes until a leaf node is reached. At every level t of the MCTS tree, if the current node C_t is corresponds to a partial solution $S = \{V_1, ..., V_k, U\}$ with t vertices already colored and k colors used, there are l possible legal moves, with $1 \leq l \leq k + 1$. Therefore, from the node C_t, l potential children $C_{t+1}^1, ..., C_{t+1}^l$ can be selected.

If $l > 1$, the selection of the most promising child node can be seen as a multi-armed bandit problem [13] with l levers. This problem of choosing the next node can be solved with the UCT algorithm for Monte Carlo tree search by selecting the child with the maximum value of the following expression [10]:

$$normalized\_score(C_{t+1}^i) + c \times \sqrt{\frac{2 * ln(nb\_visits(C_t))}{nb\_visits(C_{t+1}^i)}}, \text{ for } 1 \leq i \leq l. \quad (2)$$

Here, $nb\_visits(C)$ corresponds to the number of times the node C has been chosen to build a solution. c is a real positive coefficient allowing to balance the compromise between exploitation and exploration. It is set by default to the value of one. $normalized\_score(C_{t+1}^i)$ corresponds to a normalized score of the child node C_{t+1}^i $(1 \leq i \leq l)$ given by:

$$normalized\_score(C_{t+1}^i) = \frac{rank(C_{t+1}^i)}{\sum_{i=1}^l rank(C_{t+1}^i)},$$

where $rank(C_{t+1}^i)$ is defined as the rank between 1 and l of the nodes C_{t+1}^i obtained by sorting from bad to good according their average values $avg\_score(C_{t+1}^i)$ (nodes that seem more promising get a higher score). $avg\_score(C_{t+1}^i)$ is the mean score on the sub-branch with the node C_{t+1}^i selected obtained after all previous simulations.

3.3   Expansion

From the node C of the MCTS tree reached during the selection procedure, one new child of C is open and its corresponding legal move is applied to the current solution. Among the unopened children, we select the node associated with the lowest color number i. Therefore the child node needing the creation of a new color (and increasing the score) will be selected last.

3.4   Simulation

The simulation takes the current partial and legal solution found after the expansion phase and colors the remaining vertices. In the original MCTS algorithm, the simulation consists in choosing random moves in the set of all legal moves $\mathcal{L}(S)$ defined by Eq. (1) until the solution is completed. We call this first version *random-MCTS*. As shown in the experimental section, this version is not very efficient as the number of colors grows rapidly. Therefore, we propose two other simulation procedures:

– a constrained greedy algorithm that chooses a legal move prioritizing the moves which do not locally increase the score of the current partial solution $S = \{V_1, ..., V_k, U\}$:

$$\mathcal{L}^g(S) = \{< u, U, V_i >, 1 \leq i \leq k, \forall v \in V_i, (u, v) \notin E\} \qquad (3)$$

It only chooses the move $< u, U, V_{k+1} >$, consisting in opening a new color group and increasing the current score by $w(u)$, only if $\mathcal{L}^g(S) = \emptyset$. We call this version *greedy-random-MCTS*.

– a greedy deterministic procedure which always chooses a legal move in $\mathcal{L}(S)$ with the first available color i. We call this version *greedy-MCTS*.

3.5   Update

Once the simulation is over, a complete solution of the WVCP is obtained. If this solution is better than the best recorded solution found so far S^* (i.e., $f(S) < f(S^*)$), S becomes the new global best solution S^*.

Then, a backpropagation procedure updates each node C of the whole branch of the MCTS tree which has led to this solution:

– the running average score of each node C of the branch is updated with the score $f(S)$:

$$avg\_score(C) \leftarrow \frac{avg\_score(C) \times nb\_visits(C) + f(S)}{nb\_visits(C) + 1} \qquad (4)$$

– the counter of visits $nb\_visits(C)$ of each node of the branch is increased by one.

3.6   Pruning

During an iteration of MCTS, three pruning rules are applied:

1. during expansion: if the score $f(S)$ of the partial solution associated with a node visited during this iteration of MCTS is equal or higher to the current best-found score $f(S^*)$, then the node is deleted as the score of a partial solution cannot decrease when more vertices are colored.
2. when the best score $f(S^*)$ is found, the tree is *cleaned*. A heuristic goes through the whole tree and deletes children and possible children associated with a partial score $f(S)$ equal or superior to the best score $f(S^*)$.
3. if a node is *completely explored*, it is deleted and will not be explored in the MCTS tree anymore. A node is said *completely explored* if it is a leaf node without children, or if all of its children have already been opened once and have all been deleted. Note that this third pruning step is recursive as a node deletion can result in the deletion of its parent if it has no more children, and so on.

These three pruning rules and the fact that the symmetries are cut in the tree by restricting the set of legal moves considered at each step (see Sect. 2.2) offer the possibility to explore the whole tree in a reasonable amount of time for small instances. This peculiarity of the algorithm makes it possible to obtain an optimality proof for such instances.

4   Improving MCTS with Local Search

We now explore the possibility of improving the MCTS algorithm with local search. Coupling MCTS with a local search algorithm is motivated by the fact that after the simulation phase, the complete solution obtained can be close to a still better solution in the search space that could be discovered by local search.

For this purpose, we propose a strong integration between the MCTS framework and a local search procedure regarding two aspects:

- first, to stay consistent with the search tree learned by MCTS, we allow the local search procedure to only move the vertices of the complete solution S which are still uncolored after the selection and expansion phases. This set of free vertices is called F_V.
- secondly, the best score obtained by the local search procedure is used to update all the nodes of the branch which has led to the simulation initiation.

When coupling MCTS with a local search algorithm, the resulting heuristic can be seen as an algorithm that attempts to learn a good starting point for the local search procedure, by selecting a different path in every iteration during the selection phase.

In this work, we present the coupling of MCTS with an iterated tabu search (ITS) that searches in the same legal and partial search space presented in Sect. 2. We call this variant *MCTS+ITS*. The ITS algorithm is an adaptation of the ILS-TS algorithm [16], which is one of the best performing local search algorithms for the WVCP.

From a complete solution, the ITS algorithm iteratively performs 2 steps:

- first it deletes the heaviest vertices in F_V from 1 to 3 color groups V_i and places them in the set of uncolored vertices U.
- then, it improves the solution (i.e., minimizes the score $f(S)$) by applying the *one move* operator (consisting in moving a vertex from one color group to another) and the so-called *grenade* operator until the set of uncolored vertices U becomes empty.

The grenade operator $grenade(u, V_i)$ consists in moving a vertex u to V_i, but first, each adjacent vertice of u in V_i is relocated to other color groups or in U to keep a legal solution. In our algorithm, as we restrict the vertices allowed to move to the set of free vertices F_V, the neighborhood of a solution S associated with this grenade operator is given by: $N(S) = \{S \oplus grenade(u, V_i) : u \in F_V, u \notin V_i, 1 \leq i \leq k, \forall v \in V_i, ((u, v) \notin E \text{ or } v \in F_V)\}$.

5 Experimentation

This section aims to experimentally verify the impacts of the different greedy coloring strategies used during the MCTS simulation phase, and the interest of coupling MCTS with local search.

5.1 Experimental Settings and Benchmark Instances

A total of 161 instances in the literature were used for the experimental studies: 30 rxx graphs and 35 pxx graphs from matrix decomposition [18] and 46 small and 50 large graphs from the DIMACS and COLOR competitions.

Two preprocessing procedures were applied to reduce the graphs of the different instances. The first one comes from [22]: if the weight of a vertex of degree d is lower than the weight of the $d + 1$th heaviest vertex from any clique of the graph, then the vertex can be deleted without changing the optimal WVCP score of this instance. The second one comes from [4] and is adapted to our problem: if all neighbors of a vertex $v1$ are all neighbors with a vertex $v2$ and the weight of $v2$ is greater or equal to the weight of $v1$, then $v1$ can be deleted as it can take the color of $v2$ without impacting the score.

The MCTS algorithms of Sect. 3 were coded in C++ and compiled with the g++ 8.3 compiler and the –O3 optimization flag using GNU Parallel [21]. To solve each instance, 20 independent runs were performed on a computer equipped with an Intel Xeon E7-4850 v4 CPU with a time limit of one hour. Running the DIMACS Machine Benchmark procedure dfmax[1] on our computer took 9 s to solve the instance r500.5 using gcc 8.3 without optimization flag. The source code of our algorithm is available at https://github.com/Cyril-Grelier/gc_wvcp_mcts.

For our experimental studies, we report the following statistics in the tables of results. Column BKS (Best-Known Score) reports the best scores from all methods reported in the literature which have been obtained using different computing tools, CPU or GPU (Graphics Processing Units), sometimes under relaxed conditions (specific fine-tuning of the hyperparameters, large run time up to several days). The scores of the different methods equal to this best-known score are marked in bold in the tables. The optimal results obtained with the exact algorithm MWSS [5] and reported in [16] (launched with a time limit of ten hours for each instance) are indicated with a star. Column "best" shows the best results for a method, "Avg" the average score on the 20 runs, and "t(s)" the average time in seconds to reach the best scores.

Due to space limitations, we only present the results for 25 out of the 161 instances tested. These 25 instances can be considered to be among the most difficult ones.

---

[1] http://archive.dimacs.rutgers.edu/pub/dsj/clique/.

5.2   Monte Carlo Tree Search with Greedy Strategies

Table 1 shows the results of MCTS with the greedy heuristics for its simulation (cf. Sect. 3.4). Column 2 gives the number of vertices of the graph. Column 3 shows the best-known score ever reported in the literature for each instance. Column 4 corresponds to the baseline greedy algorithm: using the predefined vertex ordering presented in Sect. 2.3, each vertex is placed in the first color group available without creating conflicts (if no color is available, a new color group is created). Columns 5–13 show the results of the three greedy strategies used in the simulation phase of MCTS (cf. Sect. 3.4): random-MCTS, greedy-random-MCTS, and greedy-MCTS. Column 14 presents the gap between the mean score of the greedy-random-MCTS and the greedy-MCTS variants. Significant gaps (t-test with a p-value below 0.001) are underlined.

Table 1. Comparison between the random-MCTS, greedy-random-MCTS and greedy-MCTS variants (**best known**, proven optimum*, <u>significant gap</u>)

instance	$|V|$	BKS best	greedy best	random-MCTS (1) best	avg	t(s)	greedy-random-MCTS (2) best	avg	t(s)	greedy-MCTS (3) best	avg	t(s)	gap (3) - (2)
C2000.5	2000	2151	2470	3262	3281.5	3150	2584	2603.6	2524	2386	2387.7	3462	<u>-215.9</u>
C2000.9	2000	5486	6449	7300	7334.4	1128	6420	6452.3	3403	6249	6249	382.6	<u>-203.3</u>
DSJC500.1	500	184	224	240	247.6	1243	205	208.15	358	209	209	480.6	0.85
DSJC500.5	500	686	840	815	827.7	1336	760	766.5	606	757	757	1037.1	<u>-9.5</u>
DSJC500.9	500	1662	1916	1830	1857.5	680	1766	1788.3	496	1788	1788	1354.65	-0.3
DSJR500.1	500	169	187	179	184.9	656	**169**	169.05	496.05	177	177	4.6	<u>7.95</u>
flat1000_50_0	1000	924	1350	1572	1639.7	3211	1284	1304.9	3566	1273	1273.6	3516	<u>-31.3</u>
flat1000_60_0	1000	1162	1388	1666	1700.9	3571	1330	1346.75	3416	1303	1303.15	3265.26	<u>-43.6</u>
GEOM120b	120	35*	40	38	38.55	42.69	36	36.95	4	38	38	0	<u>1.05</u>
inithx.i.1	864	569	**569**	571	574.35	3584	**569**	569	0	**569**	569	0	0
latin_square_10	900	1483	1876	2178	2288.15	3575	1729	1766.25	3444	1766	1766	2746.85	-0.25
le450_25a	450	306	321	313	318.35	2311	308	313.45	3380	312	312	18.7	-1.45
le450_25b	450	307	312	310	312.7	2217.5	309	309.4	1797.85	309	309	0	-0.4
miles1500	128	797*	799	**797**	797	1.3	**797**	797	0	**797**	797	0	0
p41	116	2688*	2724	**2688**	2709.55	446.17	**2688**	2688	0.15	**2688**	2688	0	0
p42	138	2466*	2517	2489	2514.05	35	**2466**	2466.45	10	2466	2466	0.95	-0.45
queen10_10	100	162	190	173	178.2	691.5	164	168.55	1	169	169	7.35	0.45
queen16_16	256	234	264	269	274.6	854	248	252.65	116	244	244	70.25	<u>-8.65</u>
r28	288	9407*	9409	9439	9560.8	543	**9407**	9411.9	2413.8	**9407**	9407	0	<u>-4.9</u>
r29	281	8693*	8973	8875	9054	1519	**8693**	8698.05	1723.5	8694	8694	0.2	<u>-4.05</u>
r30	301	9816*	9831	9826	9881.4	770	**9816**	9819.6	27.75	9818	9818	0	-1.6
R50_5gb	50	135*	158	**135***	135	727	**135***	135	652.45	**135***	135	384.4	0
R75_1gb	70	70*	91	74	78.4	170	**70***	74.05	1150	72	72	1291.1	<u>-2.05</u>
wap01a	2368	545	628	1050	1071.95	3530	659	663.05	2278	603	603	1385.9	<u>-60.05</u>
wap02a	2464	538	619	1034	1051.2	3505	651	653.6	1268.67	591	591	1586.05	<u>-62.6</u>
			1/25	3/25			10/25			6/25			

First, we observe that the version random-MCTS (Columns 5–7) performs badly even in comparison with the baseline greedy algorithm (Column 4). This is particularly true for the largest instances such as C2000.5 and C2000.9. Indeed, for large instances, choosing random moves in the set of all legal moves is not very efficient as the number of color groups grows rapidly.

The greedy-random-MCTS variant (Columns 8–10) is equipped with a simulation procedure that chooses a legal move in priority in the set of all possible legal moves which does not locally increase the WVCP score. We observe that this simulation strategy really improves the scores of MCTS in comparison with the random-MCTS version (Columns 5–7) and can reach 10 out of the 25 best-known results. This better heuristic used during the simulation phase helped also to prove the optimality for the instance R75_1gb (indicated with a star) because the best score of 70 was found during the search which led to a pruning of the tree that allowed the tree to be fully explored earlier.

The greedy-MCTS variant (Columns 11–13) always dominates the baseline greedy algorithm (Column 3) for all the instances as it always starts with a simulation using the greedy algorithm at the root node of the tree (first simulation). It only finds 6 over 25 best-known results but is better for the largest instance when we compare the average results with those obtained by the greedy-random-MCTS variant ((Column 14). This is probably due to the fact that forcing the heaviest vertices to be grouped together in the first colors enables a better organization of the color groups. This greedy variant used during the simulation leads to a better intensification which is beneficial for large instances, for which the search tree cannot be sufficiently explored due to the time limit of one hour.

However, with the deterministic simulation of this greedy-MCTS variant, we lose the sampling part with the random legal moves used in the greedy-random-MCTS allowing greater exploration of the search space and a better estimation of the most promising branches of the search tree. This particularity of the greedy-random-MCTS allows finding the best-known score for DSJR500.1, r29, or R75_1gb for example.

To confirm this intuition, we varied the coefficient c balancing the compromise between exploration and exploitation in Eq. 2 from 0 (no exploration) to 2 (encourage exploration) for different instances. For each value of the coefficient, we performed 20 runs of greedy-random-MCTS per instance during 1h per run. Figure 1 displays 4 box plots showing the distribution of the best scores for the instances queen10_10, le450_25c, latin_square_10 and C2000.5 for the different values of c. We observe 3 typical patterns also seen for the other instances.

The pattern observed for the instance queen10_10 is typical for easy instances or medium instances. For these instances, having no exploration performs badly, then the results get better when the coefficient rises to approximately 1 where it starts to stagnate. For these instances, having no exploration rapidly leads to a local minimum trap and it seems better to secure a minimum of diversity to reach a better score.

Conversely, for very large instances such as C2000.5, we observe a pattern completely different with a best solution when $c = 0$. Indeed, for very large instances, as the search tree is huge and cannot be sufficiently explored due to the time limit, it seems more beneficial for the algorithm to favor more intensification to better search for a good solution in a small part of the tree.

An intermediate pattern appears for medium/hard instances such as le450_25c or latin_square_10 with a U-shaped curve. For these instances, an optimal value of the coefficient c is around the value of 1, with a good balance between exploitation and exploration.

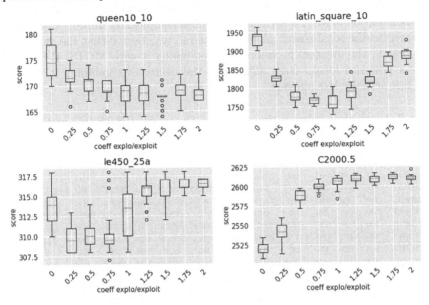

Fig. 1. Boxplots of the distribution of the scores for different values of the coefficient c between 0 and 2, for the instances queen10_10, le450_25c, latin_square_10 and C2000.5. For each configuration, 20 runs are launched with the greedy-random-MCTS variant.

5.3   Monte Carlo Tree Search with Local Search

Table 2 reports the results of the greedy-MCTS variant (Columns 4–6), the state-of-the art local search ILS-TS algorithm [16] (Columns 7–9) and the MCTS algorithm combined with ITS (MCTS+ITS) with 500 iterations per simulation (Columns 10–12). Column 13 shows the gap between the mean scores of greedy-MCTS and MCTS+ITS. Column 14 shows the gap between the mean scores of ILS-TS and MCTS+ITS. Statistically significant gaps (t-test with a p-value below 0.001) are underlined.

As shown in Table 2, the combination MCTS+ITS significantly improves the results of the greedy-MCTS version (without local search). However, the advantage of this coupling is less convincing compared with the results of the ILS-TS algorithm alone. In fact it degrades the results significantly for 2 instances DSJC500.9 and C2000.9 on average. Still, it greatly improves the average results for the flat1000_60_0 instance.

This may be due to the size of the tree that grows too fast for large instances with a high edge density such as DSJC500.9 and C2000.9. The algorithm does not allow to intensify the search enough with a timeout of 1 h. This method would probably require a much longer time to learn more information on the higher part of the tree to reach better results, in particular for large instances.

We also studied combinations of MCTS with other local searches recently proposed in the literature for this problem such as AFISA [20] or RedLS [22] (these results are available at https://github.com/Cyril-Grelier/gc_wvcp_mcts). The combination of MCTS with these methods significantly improves the results compared to the local search alone. In general, we observe that the weaker the local search, the greater the interest of combining it with MCTS. ILS-TS [16] being a powerful local search algorithm, the utility of combining it with the MCTS is actually less obvious.

Table 2. Comparison between the greedy-MCTS variant, ILS-TS [16], and MCTS + ITS. (**best known**, proven optimal*, significant gap)

instance	|V|	BKS best	greedy-MCTS (1)			ILS-TS (2)			MCTS+ITS (3)			gap (3) - (1)	gap (3) - (2)
			best	avg	t(s)	best	avg	t(s)	best	avg	t(s)		
C2000.5	2000	2151	2386	2387.7	3462	2256	2278.52	3438.27	2251	2273.45	3658	-114.25	-5.07
C2000.9	2000	5486	6249	6249	382.6	5944	5996.76	3579.44	6056	6105.4	3611	-143.6	108.64
DSJC500.1	500	184	209	209	480.6	185	188.52	3226.7	189	189.35	1485.08	-19.65	0.83
DSJC500.5	500	686	757	757	1037.1	719	735.76	320.75	721	737.9	2124	-19.1	2.14
DSJC500.9	500	1662	1788	1788	1354.65	1711	1726.1	2178.08	1728	1755.45	2900	-32.55	29.35
DSJR500.1	500	169	177	177	4.6	**169**	169	0.34	**169**	169	31.75	-8	0
flat1000_50_0	1000	924	1273	1273.6	3516	1220	1235	192.4	1211	1228.7	2555	-44.9	-6.3
flat1000_60_0	1000	1162	1303	1303.15	3265.26	1252	1271.86	283.38	1248	1261.05	2787	-42.1	-10.81
GEOM120b	120	35*	38	38	0	**35**	35	1.06	**35**	35	7.05	-3	0
inithx.i.1	864	569	**569**	569	0	**569**	569	0	**569**	569	59.45	0	0
latin_square_10	900	1483	1766	1766	2746.85	1547	1577.14	2508.73	1566	1591.1	3613	-174.9	13.96
le450_25a	450	306	312	312	18.7	**306**	306.29	679.64	**306**	306.4	1382.58	-5.6	0.11
le450_25b	450	307	309	309	0	**307**	307	51.43	**307**	307	81.25	-2	0
miles1500	128	797*	**797**	797	0	**797**	797	2.04	**797**	797	5.1	0	0
p41	116	2688*	**2688**	2688	0	**2688**	2688	0.42	**2688**	2688	4.2	0	0
p42	138	2466*	**2466**	2466	0.95	**2466**	2466	6.73	**2466**	2466	26.3	0	0
queen10_10	100	162	169	169	7.35	**162**	162	20.91	**162**	162	29.35	-7	0
queen16_16	256	234	244	244	70.25	240	241.86	873.87	241	241.85	1411	-2.15	-0.01
r28	288	9407*	**9407**	9407	0	**9407**	9407	0.31	**9407**	9407	23.2	0	0
r29	281	8693*	8694	8694	0.2	**8693**	8693	5.51	**8693**	8693	20.85	-1	0
r30	301	9816*	9818	9818	0	**9816**	9816	1.31	**9816**	9816	31	-2	0
R50_5gb	50	135*	**135***	135	384.4	**135**	135	0.21	**135**	135	1.35	0	0
R75_1gb	70	70*	72	72	1291.1	**70**	70	0.31	**70**	70	1.6	-2	0
wap01a	2368	545	603	603	1385.9	552	555.14	2499.01	550	554.7	1512	-48.3	-0.44
wap02a	2464	538	591	591	1586.05	544	546.14	1739.2	545	547.25	2867.33	-43.75	1.11
				6/25			14/25			14/25			

6   Conclusions and Discussion

In this work, we investigated Monte Carlo Tree Search applied to the weighted vertex coloring problem. We studied different greedy strategies and local searches used for the simulation phase. Our experimental results lead to three conclusions.

When the instance is large, and when a time limit is imposed, MCTS does not have the time to learn promising areas in the search space and it seems more beneficial to favor more intensification, which can be done in three different ways: (i) by lowering the coefficient which balances the compromise between exploitation and exploration during the selection phase, (ii) by using a dedicated heuristic exploiting the specificity of the problem (grouping in priority the heaviest vertices in the first groups of colors), (iii) and by using a local search procedure to improve the complete solution. For very large instances, we also observed that it may be better to use the local search alone, to avoid too early restarts with the MCTS algorithm, which prevents the local search from improving its score.

Conversely, for small instances, it seems more beneficial to encourage more exploration, to avoid getting stuck in local optima. It can be done, by increasing the coefficient, which balances the compromise between exploitation and exploration, and by using a simulation strategy with more randomness, which favors more exploration of the search tree and also allows a better evaluation of the most promising branches of the MCTS tree. For these small instances, the MCTS algorithm can provide some optimality proofs.

For medium instances, it seems important to find a good compromise between exploration and exploitation. For such instances, coupling the MCTS algorithm with a local search procedure allows finding better solutions, which cannot be reached by the MCTS algorithm or the local search alone.

Other future works could be envisaged. For instance, it could be interesting to use a more adaptive approach to trigger the local search, or to use a machine-learning algorithm to guide the search toward more promising branches of the search tree.

Acknowledgements. We would like to thank the reviewers for their useful comments. We thank Dr. Wen Sun [20], Dr. Yiyuan Wang, [22] and Pr. Bruno Nogueira [16] for sharing their codes. This work was granted access to the HPC resources of IDRIS (Grant No. 2020-A0090611887) from GENCI and the Centre Régional de Calcul Intensif des Pays de la Loire.

References

1. Brélaz, D.: New methods to color the vertices of a graph. Commun. ACM **22**(4), 251–256 (1979)
2. Browne, C.B., et al.: A survey of monte Carlo tree search methods. IEEE Trans. Comput. Intell. AI Games **4**(1), 1–43 (2012)
3. Cazenave, T., Negrevergne, B., Sikora, F.: Monte Carlo graph coloring. In: Monte Carlo Search 2020, IJCAI Workshop (2020)
4. Cheeseman, P., Kanefsky, B., Taylor, W.M.: Where the really hard problems are. In: Proceedings of the 12th International Joint Conference on Artificial Intelligence - Volume 1. IJCAI 1991, pp. 331–337. Morgan Kaufmann Publishers Inc. (1991)
5. Cornaz, D., Furini, F., Malaguti, E.: Solving vertex coloring problems as maximum weight stable set problems. Discrete Appl. Math. **217**, 151–162 (2017). https://doi.org/10.1016/j.dam.2016.09.018
6. Edelkamp, S., Greulich, C.: Solving physical traveling salesman problems with policy adaptation. In: 2014 IEEE Conference on Computational Intelligence and Games, pp. 1–8. IEEE (2014)
7. Furini, F., Malaguti, E.: Exact weighted vertex coloring via branch-and-price. Discrete Optim. **9**(2), 130–136 (2012)

8. Gelly, S., Wang, Y., Munos, R., Teytaud, O.: Modification of UCT with patterns in Monte-Carlo Go. Ph.D. thesis, INRIA (2006)
9. Goudet, O., Grelier, C., Hao, J.K.: A deep learning guided memetic framework for graph coloring problems, September 2021. arXiv:2109.05948, http://arxiv.org/abs/2109.05948
10. Jooken, J., Leyman, P., De Causmaecker, P., Wauters, T.: Exploring search space trees using an adapted version of Monte Carlo tree search for combinatorial optimization problems, November 2020, arXiv:2010.11523, http://arxiv.org/abs/2010.11523
11. Kavitha, T., Mestre, J.: Max-coloring paths: tight bounds and extensions. J. Combin. Optim. **24**(1), 1–14 (2012)
12. Kubale, M., Jackowski, B.: A generalized implicit enumeration algorithm for graph coloring. Commun. ACM **28**(4), 412–418 (1985)
13. Lai, T.L., Robbins, H.: Asymptotically efficient adaptive allocation rules. Adv. Appl. Math. **6**(1), 4–22 (1985)
14. Lewis, R.: A Guide to Graph Colouring - Algorithms and Applications. Springer, Cham (2016). https://doi.org/10.1007/978-3-319-25730-3
15. Malaguti, E., Monaci, M., Toth, P.: Models and heuristic algorithms for a weighted vertex coloring problem. J. Heuristics **15**(5), 503–526 (2009)
16. Nogueira, B., Tavares, E., Maciel, P.: Iterated local search with tabu search for the weighted vertex coloring problem. Comput. Oper. Res. **125**, 105087 (2021). https://doi.org/10.1016/j.cor.2020.105087
17. Pemmaraju, S.V., Raman, R.: Approximation algorithms for the max-coloring problem. In: Caires, L., Italiano, G.F., Monteiro, L., Palamidessi, C., Yung, M. (eds.) ICALP 2005. LNCS, vol. 3580, pp. 1064–1075. Springer, Heidelberg (2005). https://doi.org/10.1007/11523468_86
18. Prais, M., Ribeiro, C.C.: Reactive GRASP: an application to a matrix decomposition problem in TDMA traffic assignment. INFORMS J. Comput. **12**(3), 164–176 (2000)
19. Silver, D., et al.: Mastering the game of go with deep neural networks and tree search. Nature **529**(7587), 484–489 (2016). https://doi.org/10.1038/nature16961
20. Sun, W., Hao, J.K., Lai, X., Wu, Q.: Adaptive feasible and infeasible Tabu search for weighted vertex coloring. Inf. Sci. **466**, 203–219 (2018)
21. Tange, O.: GNU parallel - the command-line power tool: login. The USENIX Mag. **36**(1), 42–47 (2011)
22. Wang, Y., Cai, S., Pan, S., Li, X., Yin, M.: Reduction and local search for weighted graph coloring problem. Proc. AAAI Conf. Artif. Intell. **34**(0303), 2433–2441 (2020). https://doi.org/10.1609/aaai.v34i03.5624
23. Zhou, Y., Duval, B., Hao, J.K.: Improving probability learning based local search for graph coloring. Appl. Soft Comput. **65**, 542–553 (2018)
24. Zhou, Y., Hao, J.K., Duval, B.: Frequent pattern-based search: a case study on the quadratic assignment problem. IEEE Trans. Syst. Man Cybern. Syst. **52**, 1503–1515(2020)

A RNN-Based Hyper-heuristic
for Combinatorial Problems

Emmanuel Kieffer[1]([✉]), Gabriel Duflo[2], Grégoire Danoy[1,2], Sébastien Varrette[1],
and Pascal Bouvry[1,2]

[1] Faculty of Science, Technology and Medicine, University of Luxembourg,
Esch-sur-Alzette, Luxembourg
{emmanuel.kieffer,gregoire.danoy,sebastien.varrette,pascal.bouvry}@uni.lu
[2] Interdisciplinary Centre for Security, Reliability and Trust,
University of Luxembourg, Esch-sur-Alzette, Luxembourg
gabriel.duflo@uni.lu

Abstract. Designing efficient heuristics is a laborious and tedious task
that generally requires a full understanding and knowledge of a given
optimization problem. Hyper-heuristics have been mainly introduced to
tackle this issue and are mostly relying on Genetic Programming and its
variants. Many attempts in the literature have shown that an automatic
training mechanism for heuristic learning is possible and can challenge
human-based heuristics in terms of gap to optimality. In this work, we
introduce a novel approach based on a recent work on Deep Symbolic
Regression. We demonstrate that scoring functions can be trained using
Recurrent Neural Networks to tackle a well-know combinatorial problem,
i.e., the Multi-dimensional Knapsack. Experiments have been conducted
on instances from the OR-Library and results show that the proposed
modus operandi is an alternative and promising approach to human-
based heuristics and classical heuristic generation approaches.

Keywords: Deep Symbolic Regression · Multi-dimensional
Knapsack · Hyper-heuristics

1 Introduction

Real-word combinatorial problems are typically $\mathcal{NP}$-hard and of large size, making them intractable with exact approaches from the Operations Research literature. Numerous non-exact approaches (e.g., heuristics, metaheuristics) have thus
been proposed to provide solutions in polynomial time. Nonetheless, designing
heuristics remains a difficult exercise requiring a lot of trials and can be difficult to generalize to large scale instances. The lack of guarantees can also
be prohibitive for some decision makers. Hyper-heuristics have been therefore
designed as a methodology assisting solution designers in creating heuristics
using Evolutionary Learning. Similarly to what is done in machine learning to
build classifiers or regressive models, one can "learn to optimize" a specific problem by training a constructive model on a large set of instances. Some successful

L. Pérez Cáceres and S. Verel (Eds.): EvoCOP 2022, LNCS 13222, pp. 17–32, 2022.
https://doi.org/10.1007/978-3-031-04148-8_2

attempts in the literature (e.g. [2,4,14]) relied on constructive hyper-heuristics. The latter can be considered as a meta-algorithm that permits to engineer automatically heuristics using an existing set of instances. Hyper-heuristics essentially search through the space of heuristics or heuristic components instead of the space of solutions and use specific instances' data and properties (e.g., objective coefficients, columns for a mathematical problem) as inputs for the design of heuristics. As described in the next sections, hyper-heuristics have been historically applied to select existing heuristics and combine them together. Despite their very good results, they are constrained by the existing knowledge of a problem. This means that for a new problem with few existing heuristics, the chance to produce an efficient hyper-heuristic remains low as the search space is very restricted. On the contrary, constructive hyper-heuristic approaches assemble heuristics' components through evolution and have the advantage to spawn unseen ones. The limitations encountered by the original selective hyper-heuristics are thus removed with this constructive version. Historically, only Genetic Programming (GP) algorithms and their variants have been considered as constructive hyper-heuristics. In this work, we propose to generate heuristics and more precisely scoring functions based on a recent advance in Deep Symbolic Regression (DSR) [33]. Indeed, authors considered a Recurrent Neural Network (RNN) trained with Reinforcement Learning to provide probability distributions over symbolic expressions with the aim of solving regression problems. Learning symbolic expressions offers multiple advantages such as readability, interpretability and trustworthiness. The authors also note that the recent advances in Deep Neural Networks underexplore this aspect. We here propose to investigate the potential of this new approach to learn symbolic expressions as novel scoring functions for the Multi-dimensional Knapsack Problem (MKP). The latter has been widely study, hence our interest for it. We compare a GP-based hyper-heuristic against a RNN trained to solve MKP instances from the OR-library [12]. We demonstrate that the scoring functions obtained after training produce competitive results with state-of-the-art approaches and outperform the GP-based hyper-heuristic reference for the Multi-dimensional Knapsack.

The remainder of this article is organized as follows. The related work section details the classification of hyper-heuristics existing in the recent literature as well as the latest advances of symbolic regression. Section 3 introduces the MKP as well as some other resolution approaches. Section 4 introduces the proposed Deep Hyper-heuristic (DHH) which is subsequently compared to the reference GP-based hyper-heuristic on the MKP. Then, experimental setup and results are discussed in Sect. 5 and 6 respectively. Finally, the last section provides our conclusions and proposes some possible perspectives.

2   Related Work

Described as "heuristics to choose heuristics" by Cowling et al. [13], hyper-heuristics refer to approaches combining artificial intelligence methods to design heuristics. Contrary to algorithms searching in the space of solutions, hyper-heuristic algorithms search in the space of algorithms, i.e. heuristics, in order to determine the best heuristics combination to solve a problem. Burke et al. in [8] compared hyper-heuristics as *"off-the-peg"* methods which are generic approaches providing solutions of acceptable quality as opposed to *"made-to-measure"* techniques. This need of generalization is clearly related to machine learning approaches. Therefore, hyper-heuristics can be classified as learning algorithms and have been motivated by the following factors: the difficulty of maintaining problem-specific algorithms and the need of automating the design of algorithms. Two methodologies of hyper-heuristics rose from the literature: the first one is referred to as *heuristic selection* while the second one is described as *heuristic generation*.

Heuristic selection is the "legacy" approach which involves determining the best subset of heuristics solving a problem. Among these approaches, we can distinguish constructive and perturbation methods. Constructive methods assemble a solution step by step, starting from a partial or empty solution. The construction of a full solution is achieved through the selection and application of a heuristic to this partial solution. For this purpose, a pre-existing set of heuristics should be provided in order to determine the best heuristics to apply at a given state of the search. The resolution then stops when the solution is complete. Constructive methods have been applied for instance on vehicle routing [23], 2D packing [28], constraint satisfaction [32] and scheduling [22]. On the contrary, perturbation methods start from a valid solution and attempt to modify it using a pre-existing set of perturbation heuristics. At each step, one heuristic is selected from this set and applied to the solution. According to a specific acceptance strategy factor (e.g., deterministic or non-deterministic), the new solution is accepted or rejected. It is also possible to perturb multiple solutions at once but it has been seldom used in the literature. Scheduling [25], space allocation [5] and packing [6] are problems where such perturbation methods have been exploited.

More recently, a growing interest has been devoted to heuristic generation. The motivation behind this approach is the automatic generation mechanism which does not rely on a possible set of pre-existing heuristics. Instead of searching in the space of heuristics, the hyper-heuristic searches in the space of components, i.e., instance data. Building a complete heuristic is not a trivial task but it has been performed using Genetic Programming. In contrast to Genetic Algorithms (GA) where solution vectors are improved via genetic operators, GP algorithms evolve a population of programs until a certain stopping criterion is satisfied. Programs are expressed as tree structures which means that their length is not defined a priori contrary to GAs. The suitability of GP algorithms to produce heuristics has been outlined in a survey by Burke et al. [9]. The major advantage brought by GP algorithms is the possibility to automatize the assembly of building blocks, i.e. terminal sets and function sets emerging from knowledge gained on a problem. Concerning the MKP, this knowledge can be easily retrieved from the literature. Additionally, the dynamic length of the tree encoding is an advantage if some size limitations are implemented. Indeed, large programs will tend to have over-fitting symptoms meaning that the generated heuristics will be very efficient on the training instances but not on new ones. These are typically the same issues faced by machine learning models. GP-based hyper-heuristics encountered real successes in cutting and packing [10], function optimization [31] and other additional domains [18,30,40]. In addition, it is worth mentioning the recent approaches such as Cartesian GP and Grammar-based GP algorithms which are improvements of classical GP algorithms. Cartesian GP algorithms is an alternative form of GP algorithms encoding a graph representation of a computer program. Cartesian GP defines explicitly a size preventing bloat but can be very sensitive to parameters. In Grammar-based GP algorithms [7], a grammar in Backus-Naur Form (BNF) is considered to map linear genotypes to phenotype trees and have less structural difficulties than a classical GP algorithms.

Contrary to [33] which only considers regression problems, we extend the field of application of Deep Recurrent Networks to the task of learning scoring functions for a combinatorial problem such as the MKP. Nonetheless, it is worth mentioning recent advances in symbolic regression using deep neural networks although there are few attempts in the literature. For instance, a neural network implementation has been investigated in [39] as a pre-processing approach before using symbolic regression. In [36], authors have explored symbolic operators as activation functions while keeping neural networks differentiable. In [26], variational encoders have been considered for the first time to encode and decode parse trees using predefined grammar rules. Finally, [33] recently proposed a RNN to generate probability distributions over symbolic expressions. Authors have relied on pre-order traversal to build abstract syntax tree representing equations. They also illustrate and compare their approach with other frameworks and outperformed most of them. This is the reason why we rely on this last contribution to build a new type of hyper-heuristic. One should also note that the aforementioned works only investigated small scale regression problems and their suitability still needs to be demonstrated for large-scale problems. In this work, we partially answer this last question when extending the approach to hyper-heuristics.

3 Multi-dimensional 0-1 Knapsack

The Multi-dimensional 0-1 Knapsack (MKP) is a $\mathcal{NP}$-hard combinatorial problem which extends the well-know 0-1 Knapsack problem for multiple sacks. The objective is to find a subset of items maximizing the total profit and fitting into the m sacks. Each item j gives a profit p_j and occupies some space a_{ij} in the sack i. Each sack i has a maximum capacity b_i which should not be exceeded. The Multi-dimensional 0-1 Knapsack can be formally expressed as a 0-1 Integer Linear Program (ILP) as illustrated by Program 1.

More practically, the MKP is a resource allocation problem. It received a wide attention from many communities, including the Operations Research and Evolutionary Computing ones. Multiple heuristics and metaheuristics have been designed to tackle the MKP in addition to the existing exact approaches which

$$\text{maximize} \quad \sum_{j=1}^{n} p_j x_j \tag{1}$$

$$\text{subject to} \quad \sum_{j=1}^{n} a_{ij} x_j \leq b_i \quad \forall i \in \{1, ..., m\} \tag{2}$$

$$x_j \in \{0, 1\} \quad \forall j \in \{1, ..., n\} \tag{3}$$

Program. 1. 0-1 ILP for the multi-dimensional knapsack

can only handle small instances. Among existing heuristics for the MKP, *greedy* heuristics are designed to be fast, i.e., work in polynomial time. Generally, these are constructive methods that can be categorized as primal or dual heuristics. A primal heuristic starts from a feasible solution and tries to improve the objective value while keeping the solution feasible. On the contrary, a dual heuristic starts from an upper-bound solution (in case of maximization), i.e., not feasible, and attempts to make it feasible while minimizing the impact on the objective value. For example, [37] considered a dual heuristic with a starting solution taking all items. Then, the heuristic removed items according to an increasing ratio until feasibility was reached. The ratio or score of each item j is computed as follows: $r_j = \frac{p_j}{\sum_i^m w_i a_{ij}}$. Weights w_i are sometimes omitted since they add a new level of complexity and are specific to the considered instances. Using Lagrangian relaxation, [29] improves the dual heuristic of [37].

Concerning primal heuristics, items are added as long as all constraints remain satisfied. In this case, items with the largest ratios have priority. These new heuristics using dual multipliers give insights about the variables to fix. Further improvements based on bound tightness [21], threshold acceptance [17] and noising approaches [11] have contributed to improve such heuristics. The interested reader may refer to the survey on MKP heuristics by Fréville [20].

Metaheuristics have also been considered to solve MKP instances. These are stochastic algorithms which successfully tackled many combinatorial optimization problems, including the MKP. A simulated annealing algorithm has been first employed in [16] where specific random moves should maintain feasibility during the search. Many diverse metaheuristics have been then considered to solve the MKP during the last decade including Ant Colony Optimization [19], Genetic algorithms [27], Memetic algorithms [12], Particle Swarm algorithms [24], Fish Swarm algorithms [3] and Bee Colony algorithms [38].

4   A RNN-Based Hyper-heuristic

Contrary to most GP-based hyper-heuristics building full syntax trees and applying evolutionary operators to generate new ones, we propose hereafter to consider RNNs for this task. Based on the RNN architecture proposed in [33], we posit that Deep Recurrent Networks for symbolic expressions are perfectly suitable to learn *scoring functions* for a combinatorial problem such as the MKP. A scoring function takes as inputs information about an item to be added to the sacks.

This function measures the pertinence of the given item j which is represented by the profit p_j and the column of the constraint matrix $A_{.j}$.

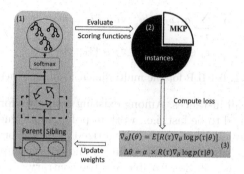

Fig. 1. Workflow of the RNN-based Hyper-Heuristic (RHH): sampling (1), evaluation (2) and training (3).

Although the DSR approach proposed in [33] has the unique purpose to tackle symbolic regression problems, we propose to extend it in order to create a RNN-based Hyper-heuristic (RHH) generating sequences of tokens described in Table 1. A sequence can then be decoded into a symbolic expression represented as a binary tree, i.e., each node can only have at most one sibling. The sequence of tokens is produced by the RNN in a autoregressive manner, i.e., k^{th} token prediction depends on the previously obtained tokens. RNNs model efficiently time-varying information such as sequences. The RHH implementation presented hereafter provides a one-to-one mapping between the sampled symbolic expressions and the resulting scoring functions to solve MKP instances. This implementation can be decomposed into 3 main iterative steps as depicted in Fig. 1. The first one (1) is the sampling step in which a batch of symbolic expressions is produced. The second step (2) turns symbolic expressions into scoring functions which are subsequently evaluated using a greedy heuristic template to generate rewards. Finally, the last step (3) performs a one-step gradient ascent to update the embedded RNN's weights using Policy Gradient. All these steps are described in details in the following sections.

4.1   Sampling Symbolic Expressions (1)

Let us define T, the set of tokens which can be sampled. This set is equivalent to the union of terminal and non-terminal sets defined in GP algorithms. Table 1 provides a description of all tokens considered to generate scoring functions for the Multi-dimensional Knapsack. Among these tokens, we selected the entire column describing an item, the average difference between the capacity and the average resource consumption for item j, the maximal resources consumption and the total resource consumption of item j. In order to discriminate **good** items from **bad** ones, we also add as prior knowledge the solution of the LP relaxation which is clearly a very good feature for our learning purpose. The set shown in Table 1 is not exhaustive and could be easily completed with new features to discriminate more accurately profitable items from valueless ones.

Table 1. The set of tokens T which are components of scoring functions

Name	Description
Operators	
+	Add two inputs
-	Substract two inputs
*	Multiply two inputs
%	Divide two inputs with protection
Terminal sets/Arguments	
p_j	Profit of the current item j
$d_j = \frac{\sum_i b_i - a_{ij}}{m}$	Average difference between the capacity and the resource consumption for item j
a_{1j}	Resource consumption of item j for sack 1
a_{2j}	Resource consumption of item j for sack 1
$a_{\dots,j}$	...
a_{mj}	Resource consumption of item j for sack m
$s_j = \sum_i a_{ij}$	Total resource consumption of item j for sack i
$m_j = \max_i a_{ij}$	Max resource consumption of item j for sack i
$\bar{x}_j$	Solution value for item j after LP relaxation

The following describes how scoring functions are created from sampled symbolic expressions using a Recurrent Neural Network (RNN) as illustrated in Fig. 2. The first inputs provided to the networks are necessarily empty since the tree representation of the future symbolic expression is empty (see **step 0** in Fig. 2). Therefore, an empty token $<E>$ is added to the set of tokens listed in Table 1. At each step i, the next token τ_i^s is sampled according to the pre-order traversal. Its future location in the tree is thus known which means that the parent and sibling nodes can be identified and provided as inputs to the RNN cell. This is illustrated in Fig. 2 where the left-child of the root node will host the next token. Its parent is obviously the root node and it has no sibling yet. The RNN returns a logit vector L_i which is passed to a softmax layer. This very common layer permits to obtain a discrete probability distribution over all tokens at step i, i.e., $\sum_{k=1}^{|T|} p(\tau_i^k | \tau_{i-1}^s; \theta) = 1$. The sampled token τ_i^s ($\bar{x}_j$ in step 1) is then added to the tree. Please note that sampled tokens τ_i^s and τ_{i-1}^s are not necessarily connected in the resulting tree. In fact, they can be far from each other as depicted in Fig. 2. This is due to the hierarchical structure induced by the traversal. As a consequence, a sample token at step $i - 1$ is not necessarily provided as input for step i. This is illustrated by **step 5** in Fig. 2. "Teacher forcing" is therefore considered to replace inputs by the true parent and sibling with regards to the traversal. Finally when the expression is complete, i.e., no empty leaves, one can define the log-likelihood of the sample sequence of tokens τ, i.e., sampled symbolic expression, as follows: $\log p(\tau|\theta) = \sum_{i=1}^{|\tau|} \log p(\tau_i^s|\tau_{i-1}^s;\theta)$.

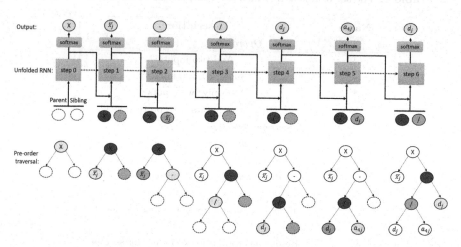

Fig. 2. Example of scoring function: $\bar{x}_j(\frac{d_j}{a_j^4} - d_j)$ generated from a sampled sequence of tokens (length $= 7$). Blue circles represent parent inputs while green ones stand for sibling. Outputs are illustrated by yellow circles and are obtained after applying a softmax layer and sampling with regards to a discrete probability distribution over all possible tokens. Dashed circles represent missing token positions still need to be filled. At each step, parent and sibling tokens are presented to the RNN cell. The tree representation grows according to the pre-order traversal (recursively traverse left subtree first) (Color figure online)

4.2   Evaluation of the Resulting Scoring Functions (2)

A sampled symbolic expression can then be turned into a scoring function applied inside a so-called heuristic template (see Algorithm 1) in order to evaluate its relevance to provide an efficient insertion order of items into the sacks. The combination of this template and a scoring function characterizes a heuristic which is subsequently applied on a set of multiple MKP instances I, i.e., a training set. In this work, we consider a primal heuristic template starting from a feasible and trivial solution and selecting items ranked using a generated scoring function until sacks are full. Ranking is obtained by applying the scoring function on item data.

4.3   RNN Training and Gradient Update (3)

In order to train the RNN to produce efficient scoring functions, a reward/fitness for the sequence of tokens $R(\tau)$ is proposed as follows: $R(\tau) = \frac{1}{1+\sum_{j=1}^{|I|} v_\tau^j}$ with

Algorithm 1. greedy_heuristic(instance,function)

```
1: value ← 0
2: solution ← [0,0,...,0]
3: sacks ← [0,0,...,0]
4: indexes ← sort(items,function)
5: while indexes ≠ ∅ do
6:     index ← indexes.pop_head()
7:     if sacks[i] + instance.A[i, index] ≤ instance.rhs[i] ∀i ∈ {1, ..., m} then
8:         solution[index] ← 1
9:         value ← value + instance.p[index]
10:        for i ∈ {1, ..., m} do
11:            sacks[i] ← sacks[i] + instance.A[i, index]
12:        end for
13:    end if
14: end while
15: return value
```

v_τ^j the solution value obtained by solving the jth instance of I with the scoring function generated from the sequence of tokens τ.

The RNN is trained on batches of N sampled expressions $\mathcal{B} = \{\tau^{(k)}\}_{k=1}^N$ using **Policy Gradients** $\nabla_\theta J(\theta) = E_\tau [R(\tau)\nabla_\theta \log p(\tau|\theta)]$. The loss function $\mathcal{L}(\theta)$ is therefore expressed in this way:

$$\mathcal{L}(\theta) = \frac{1}{N} \sum_{k=1}^N R(\tau^{(k)}) \log p(\tau^{(k)}|\theta) \tag{4}$$

Although solving MKP instances is deterministic, the spread of rewards can be very large. Consequently, the policy gradient will have high-variance. One way to mitigate this problem is to substract a baseline to rewards by some constant value, which is normally the mean of the rewards. The RNN training steps are described hereafter:

1. Initialise the network with random weights
2. Generate a batch $\mathcal{B}$ of N sampled expressions
3. Expressions are turned into scoring functions and injected into the heuristic template (see Algorithm 1)
4. Heuristics are applied on a set $\mathcal{I}$ of training instances and rewards $R(\tau^k)$ are computed based on the resolution of instances.
5. The $\mathcal{L}(\theta) = \frac{1}{N} \sum_{k=1}^N R(\tau^{(k)}) \log p(\tau^{(k)}|\theta)$ is computed according to rewards and the log-likelihood
6. Perform a gradient ascent step update of the network's weights, maximizing the policy gradient
7. Repeat 2. Until stopping criterion is met

The next section defines experimental setup to assess the proposed hyper-heuristic, namely RHH.

5   Experimental Setup

The MKP instances from the OR-Library[1] have been considered. These have been originally introduced by [12] and include 270 instances classified according to the number of variables(items) $n \in \{100, 250, 500\}$, number of constraints $m \in \{5, 10, 30\}$ and tightness ratio $r \in \{0.25, 0.50, 0.75\}$.

In order to evaluate the efficiency of the trained heuristics, i.e., the combination of the heuristic template and trained scoring functions, on these instances, we adopt as performance measure the %-gap (see Eq. 5) between a lower bound and an upper bound. Lower bounds are provided by the heuristics, i.e., $\underline{v}^h$, while the continuous LP relaxation, i.e., $\overline{v}_{lp}$, will be the reference upper bound. In addition, we multiply all gaps by 100.

$$\%\text{-}gap = 100 * \frac{\overline{v}_{lp} - \underline{v}^h}{\overline{v}_{lp}} \tag{5}$$

The RNN-based Hyper-heuristic (RHH) will be compared to a GP-based hyper-heuristic described in [14]. Table 2 details all the GP parameters and GP operators used by these authors. They performed 50 generations with a population size of 10000 scoring functions. Contrary to GAs, GP algorithms make a different use of the evolutionary operators. First of all, their probabilities should sum to 1. For example, in the case of Table 2, 85% of the solutions will mate with another one, 10% will face mutations and only 5% will be kept without any modifications for the next generation. In order to keep control of the size of each syntax tree, they prevent trees from having a depth greater than 17 nodes. The interested reader can refer to [14] for more details on the GP implementation.

Table 2. GP parameters

Generations	50
Population size	10000
Crossover Probability (CXPB)	0.85
Mutation Probability (MUTPB)	0.1
Reproduction Probability	0.05
Tree initialization method	Ramped half-and-half
Selection Method	Tournament selection with size = 7
Depth limitation	17
Crossover Operator	One point crossover
Mutation Operator	Grow

Finally both approaches, i.e., RHH and the GP-based hyper-heuristic, follow the same protocol to train scoring functions. All instances have been divided

---

[1] http://people.brunel.ac.uk/mastjjb/jeb/orlib/mknapinfo.html.

into groups depending on the number of variables, the number of constraints and the tightness ratio. Both hyper-heuristics have been applied on all groups which contain ten instances each. Five random instances have been selected as training instances while the remaining five instances have been considered as validation instances. The reported %-gaps are only computed on the validation instances. For each group, five runs have been performed in order to obtain an average %-gap and the best scoring function has been recorded.

Table 3 lists all hyperparameters describing the RNN. A single layer with 32 units has been considered. A maximum of 500000 sampled expressions are generated to provide fair comparisons with the GP-based hyper-heuristic. The RNN is trained on batches of 1000 sampled expressions. The Adam optimiser is set up with a learning rate of 0.0005 and has been selected to optimise the RNN's weights.

Table 3. RNN parameters

Max sample	500000
Batch size	1000
RNN cell type	LSTM
Number of layers	1
Number of units	32
Optimizer	Adam
Learning rate	0.0005
Hidden state initializer	zeros

Experiments have been conducted on the High Performance Computing (HPC) platform of the University of Luxembourg. Each run was completed on a single core of an Intel Xeon E5-2680 v3 @ 2.5 GHz, 32 GB of RAM server, which was dedicated to this task.

6   Experimental Results

The average %-gap obtained after 5 runs is provided in Table 4. The left part of this table represents the results reported in [14] while the right part corresponds to the RNN-based Hyper-Heuristic approach (RHH) proposed in this work.

Each row depicts a specific instance set **ORnXm** divided into groups of different tightness ratios. For example, the average %-gap obtained for the instance set **OR5x100** with tightness ratio 0.25 is **4.98** for the GP-based Hyper-heuristic approach. Gray shaded cells indicate that the average %-gap is better for the considered approach. For example, the average %-gap obtained for the instance set **OR5x100** with tightness ratio 0.25 is reported better for the RHH approach.

Table 4 shows us that each instance set **ORnXm** has a better average %-gap when solved with the RHH approach. When considering tightness ratios, we can observe that RHH outperforms all instances with $r = 0.25$ and $r = 0.5$ while this is not the case for $r = 0.75$. The tightness ratio defines the scarcity of capacities. The closer to 0 the tightness ratio the more constrained the instance. Indeed, a ratio $r = 0.25$ implies that about 25% of the items can be packed contrary to a ratio $r = 0.75$ where 75% of the items can be packed. These results show that the proposed RHH approach is able to handle more efficiently different levels of tightness.

Table 4. Average gaps (%) of the best found heuristics on the ORlib instances ordered by tightness ratio

Instance set	Original approach				RHH			
	0.25	0.50	0.75	Average	0.25	0.50	0.75	Average
OR5x100	4.98	2.05	1.36	*2.80*	1.58	2.02	2.33	*1.98*
OR5x250	3.08	1.66	0.77	*1.84*	0.49	0.47	0.85	*0.60*
OR5x500	2.38	1.64	0.71	*1.58*	0.26	0.22	0.51	*0.33*
OR10x100	7.39	3.54	2.26	*4.40*	2.30	2.34	3.38	*2.67*
OR10x250	4.43	2.78	1.15	*2.79*	1.01	0.66	1.25	*0.98*
OR10x500	3.77	1.97	0.99	*2.24*	0.38	0.35	0.57	*0.43*
OR30x100	8.67	4.70	2.43	*5.27*	3.41	2.06	6.01	*3.83*
OR30x250	5.73	3.25	1.70	*3.56*	1.33	1.20	2.34	*1.62*
OR30x500	4.80	2.54	1.40	*2.91*	0.80	0.56	0.95	*0.77*
All instances	**5.03**	**2.68**	**1.42**	**3.04**	**1.29**	**1.10**	**2.02**	**1.47**

The best scoring functions obtained with RHH are listed in Table 5. One can notice the presence of $\bar{x}_j$, i.e., the solution of the LP relaxation for variable x_j, in all resulting scoring functions. Interestingly, multiple scoring functions do not include the profit p_j (e.g., OR30x500-0.75). The size of each scoring function is rather reasonable and no "bloating" effect, generally experienced with GP algorithms, can be observed.

Table 5. Best scoring functions obtained for each benchmark

Instance set	Scoring funtions		
	0.25	0.50	0.75
OR5x100	$\bar{x}_j - (d_j + s_j) * \bar{x}_j^2/p_j$	$((-d_j * s_j * \bar{x}_j + 1)/s_j - a_{1j})/(a_{3j} + m_j)$	$((a_{1j} + d_j)/a_{2j} - d_j) * \bar{x}_j$
OR5x250	$\bar{x}_j/(a_{1j} * \bar{x}_j - s_j)$	$\bar{x}_j/(-(a_{5j} + d_j)/m_j - a_{1j} - s_j)$	$-(-(a_{3j} - m_j * \bar{x}_j)/a_{1j} + d_j - s_j) * \bar{x}_j + d_j$
OR5x500	$-p_j - d_j * \bar{x}_j^2 + d_j$	$-(p_j + s_j * \bar{x}_j^2 - \bar{x}_j) * \bar{x}_j + a_{1j}$	$(-(d_j + d_j/p_j) * \bar{x}_j + a_{5j} + s_j + \bar{x}_j)/(s_j^2 + \bar{x}_j)$
OR10x100	$(a_{1j} - s_j) * \bar{x}_j$	$(-(p_j + a_{2j})) * \bar{x}_j + a_{2j}) * \bar{x}_j$	$((p_j + a_{9j})/s_j) - \bar{x}_j$
OR10x250	$-\bar{x}_j + 1/a_{1j}$	$(a_{10j}/s_j) - \bar{x}_j$	$(a_{2j} - a_{4j} - s_j)) * \bar{x}_j^2$
OR10x500	$(p_j - s_j) * \bar{x}_j^2$	$-p_j * d_j * \bar{x}_j + a_{4j}$	$\bar{x}_j/(-(a_{2j} + s_j) * d_j + a_{4j} + 2a_{5j})$
OR30x100	$p_j * \bar{x}_j/(-s_j + \bar{x}_j)$	$\bar{x}_j/(-a_{27j} - d_j + \bar{x}_j)$	$-p_j * \bar{x}_j + d_j$
OR30x250	$a_{12j} - s_j * \bar{x}_j$	$(-p_j + \bar{x}_j) * \bar{x}_j$	$p_j * (a_{24j} + a_{29j} - s_j) * \bar{x}_j$
OR30x500	$p_j * \bar{x}_j/(-d_j + s_j)$	$\bar{x}_j/(a_{7j} - s_j)$	$\bar{x}_j/(-d_j + s_j)$

Table 6. Comparisons with multiple existing approaches over all ORlib instances in terms of gap (%)

Type	Reference	%-gap
MIP	[15] (CPLEX 12.2)	0.52
MA	[12]	0.54
Selection HH	[15]	0.70
MA	[42]	0.92
Heuristic	[34]	1.37
RHH	**this work**	**1.47**
Heuristic	[21]	1.91
Metaheuristic	[35]	2.28
GHH	[14]	3.04
MIP	[12] (CPLEX 4.0)	3.14
Heuristic	[1]	3.46
Heuristic	[41]	6.98
Heuristic	[29]	7.69

Last but not least, Table 6 presents the %-gap obtained by different existing methods from the literature on the same benchmarks. The approach proposed in this work, i.e., RHH, obtains a good rank, i.e., 6th position which demonstrates the suitability of using recurrent neural architectures to assist in building heuristics.

Hyper-heuristics or automatic generation of heuristics are not dedicated to provide the best results. They are general approaches which have been proposed to facilitate the generation of **good performing** and **fast** algorithms to solve problems. Table 6 shows that despite their general approach, they can provide better results than dedicated algorithms.

7   Conclusion and Perspectives

Traditionally, hyper-heuristics are GP-based approaches evolving heuristics represented by abstract syntax trees. In this paper, we proposed a new hyper-heuristic model based on deep symbolic expressions to automatically solve combinatorial problems such as the Multidimensional Knapsack. We tackled the generation of scoring functions to measure the pertinence of adding an item to the sacks. These functions therefore allow finding an inserting order in the sacks and provide reasonable educated guesses to solve the MKP. Contrary to the classical knapsack with a single constraint, it is not trivial to manually discover an efficient scoring procedure for multi-dimensional variants of the knapsack. After detailing the methodology, we compared a state-of-the-art GP hyper-heuristic versus the new deep hyper-heuristic approach. To measure the performance of both approaches and fairly confront them, validation instances which have been not presented to both hyper-heuristics during training served to compute a performance measure, i.e., %-gap. Results show that the proposed methodology relying on this recurrent neural architecture outperforms the classical GP-based hyper-heuristic on this problem. Training symbolic expressions with deep learning has the benefit to provide efficient predictions while keeping scoring functions explainable. Symbolic expressions can be easily analysed by experts contrary to a network providing only black-box scoring values which would be difficult to interpret. Future works will attempt to apply the proposed approach to "Column Generation (CG)", a well-known Operation Research technique, to solve large-scale problems. This would notably be helpful to cope with degeneracy.

References

1. Akçay, Y., Li, H., Xu, S.H.: Greedy algorithm for the general multidimensional knapsack problem. Ann. Oper. Res. **150**(1), 17–29 (2006). https://doi.org/10.1007/s10479-006-0150-4
2. Allen, S., Burke, E.K., Hyde, M., Kendall, G.: Evolving reusable 3D packing heuristics with genetic programming. In: Proceedings of the 11th Annual Conference on Genetic and Evolutionary Computation - GECCO 2009. ACM Press (2009). https://doi.org/10.1145/1569901.1570029
3. Azad, M.A.K., Rocha, A.M.A.C., Fernandes, E.M.G.P.: Solving large 0–1 multidimensional knapsack problems by a new simplified binary artificial fish swarm algorithm. J. Math. Model. Algorithms Oper. Res. **14**(3), 313–330 (2015). https://doi.org/10.1007/s10852-015-9275-2
4. Bader-El-Den, M., Poli, R.: Generating SAT local-search heuristics using a GP hyper-heuristic framework. In: Monmarché, N., Talbi, E.-G., Collet, P., Schoenauer, M., Lutton, E. (eds.) EA 2007. LNCS, vol. 4926, pp. 37–49. Springer, Heidelberg (2008). https://doi.org/10.1007/978-3-540-79305-2_4
5. Bai, R., Burke, E.K., Kendall, G.: Heuristic, meta-heuristic and hyper-heuristic approaches for fresh produce inventory control and shelf space allocation. J. Oper. Res. Soc. **59**(10), 1387–1397 (2008). https://doi.org/10.1057/palgrave.jors.2602463
6. Bai, R., Blazewicz, J., Burke, E.K., Kendall, G., McCollum, B.: A simulated annealing hyper-heuristic methodology for flexible decision support. 4OR-Q J. Oper. Res. **10**(1), 43–66 (2011). https://doi.org/10.1007/s10288-011-0182-8
7. Brabazon, A., O'Neill, M., McGarraghy, S.: Grammar-based and developmental genetic programming. In: Natural Computing Algorithms. NCS, pp. 345–356. Springer, Heidelberg (2015). https://doi.org/10.1007/978-3-662-43631-8_18
8. Burke, E.K., et al.: Hyper-heuristics: a survey of the state of the art. J. Oper. Res. Soc. **64**(12), 1695–1724 (2013). https://doi.org/10.1057/jors.2013.71
9. Burke, E.K., Hyde, M.R., Kendall, G., Ochoa, G., Ozcan, E., Woodward, J.R.: Exploring hyper-heuristic methodologies with genetic programming. In: Mumford, C.L., Jain, L.C. (eds.) Computational Intelligence. ISRL, vol. 1, pp. 177–201. Springer, Heidelberg (2009). https://doi.org/10.1007/978-3-642-01799-5_6
10. Burke, E.K., Hyde, M.R., Kendall, G.: Grammatical evolution of local search heuristics. IEEE Trans. Evol. Comput. **16**(3), 406–417 (2012). https://doi.org/10.1109/tevc.2011.2160401

11. Charon, I., Hudry, O.: The noising method: a new method for combinatorial optimization. Oper. Res. Lett. **14**(3), 133–137 (1993). https://doi.org/10.1016/0167-6377(93)90023-a

12. Chu, P., Beasley, J.: A genetic algorithm for the multidimensional knapsack problem. J. Heuristics **4**(1), 63–86 (1998). https://doi.org/10.1023/A:1009642405419

13. Cowling, P., Kendall, G., Soubeiga, E.: A hyperheuristic approach to scheduling a sales summit. In: Burke, E., Erben, W. (eds.) PATAT 2000. LNCS, vol. 2079, pp. 176–190. Springer, Heidelberg (2001). https://doi.org/10.1007/3-540-44629-x_11

14. Drake, J.H., Hyde, M., Khaled, I., Özcan, E.: A genetic programming hyperheuristic for the multidimensional knapsack problem. Kybernetes **43**(9/10), 1500–1511 (2014). https://doi.org/10.1108/k-09-2013-0201

15. Drake, J.H., Özcan, E., Burke, E.K.: A case study of controlling crossover in a selection hyper-heuristic framework using the multidimensional knapsack problem. Evol. Comput. **24**(1), 113–141 (2016). https://doi.org/10.1162/evco_a_00145

16. Drexl, A.: A simulated annealing approach to the multiconstraint zero-one knapsack problem. Computing **40**(1), 1–8 (1988). https://doi.org/10.1007/bf02242185

17. Dueck, G., Scheuer, T.: Threshold accepting: a general purpose optimization algorithm appearing superior to simulated annealing. J. Comput. Phys. **90**(1), 161–175 (1990). https://doi.org/10.1016/0021-9991(90)90201-b

18. Elyasaf, A., Hauptman, A., Sipper, M.: Evolutionary design of FreeCell solvers. IEEE Trans. Comput. Intell. AI Games **4**(4), 270–281 (2012). https://doi.org/10.1109/tciaig.2012.2210423

19. Fingler, H., Cáceres, E.N., Mongelli, H., Song, S.W.: A CUDA based solution to the multidimensional knapsack problem using the ant colony optimization. Procedia Comput. Sci. **29**, 84–94 (2014). https://doi.org/10.1016/j.procs.2014.05.008

20. Fréville, A.: The multidimensional 0–1 knapsack problem: an overview. Eur. J. Oper. Res. **155**(1), 1–21 (2004). https://doi.org/10.1016/s0377-2217(03)00274-1

21. Freville, A., Plateau, G.: An efficient preprocessing procedure for the multidimensional 0–1 knapsack problem. Discrete Appl. Math. **49**(1–3), 189–212 (1994). https://doi.org/10.1016/0166-218x(94)90209-7

22. García-Villoria, A., Salhi, S., Corominas, A., Pastor, R.: Hyper-heuristic approaches for the response time variability problem. Eur. J. Oper. Res. **211**(1), 160–169 (2011). https://doi.org/10.1016/j.ejor.2010.12.005

23. Garrido, P., Riff, M.C.: DVRP: a hard dynamic combinatorial optimisation problem tackled by an evolutionary hyper-heuristic. J. Heuristics **16**(6), 795–834 (2010). https://doi.org/10.1007/s10732-010-9126-2

24. Hembecker, F., Lopes, H.S., Godoy, W.: Particle swarm optimization for the multidimensional knapsack problem. In: Beliczynski, B., Dzielinski, A., Iwanowski, M., Ribeiro, B. (eds.) ICANNGA 2007. LNCS, vol. 4431, pp. 358–365. Springer, Heidelberg (2007). https://doi.org/10.1007/978-3-540-71618-1_40

25. Kendall, G.: Scheduling English football fixtures over holiday periods. J. Oper. Res. Soc. **59**(6), 743–755 (2008). https://doi.org/10.1057/palgrave.jors.2602382

26. Kusner, M.J., Paige, B., Hernández-Lobato, J.M.: Grammar variational autoencoder. In: Precup, D., Teh, Y.W. (eds.) Proceedings of the 34th International Conference on Machine Learning. Proceedings of Machine Learning Research, vol. 70, pp. 1945–1954. PMLR, 06–11 August 2017

27. Lai, G., Yuan, D., Yang, S.: A new hybrid combinatorial genetic algorithm for multidimensional knapsack problems. J. Supercomput. **70**(2), 930–945 (2014). https://doi.org/10.1007/s11227-014-1268-9

28. López-Camacho, E., Terashima-Marín, H., Ross, P., Valenzuela-Rendón, M.: Problem-state representations in a hyper-heuristic approach for the 2D irregular BPP. In: Proceedings of the 12th Annual Conference on Genetic and Evolutionary Computation - GECCO 2010. ACM Press (2010). https://doi.org/10.1145/1830483.1830539

29. Magazine, M., Oguz, O.: A heuristic algorithm for the multidimensional zero-one knapsack problem. Eur. J. Oper. Res. **16**(3), 319–326 (1984). https://doi.org/10.1016/0377-2217(84)90286-8

30. Nguyen, S., Zhang, M., Johnston, M.: A genetic programming based hyper-heuristic approach for combinatorial optimisation. In: Proceedings of the 13th Annual Conference on Genetic and Evolutionary Computation, GECCO 2011, pp. 1299–1306. ACM, New York (2011). https://doi.org/10.1145/2001576.2001752

31. Oltean, M.: Evolving evolutionary algorithms using linear genetic programming. Evol. Comput. **13**(3), 387–410 (2005). https://doi.org/10.1162/1063656054794815

32. Ortiz-Bayliss, J.C., Ozcan, E., Parkes, A.J., Terashima-Marin, H.: Mapping the performance of heuristics for constraint satisfaction. In: IEEE Congress on Evolutionary Computation. IEEE, July 2010. https://doi.org/10.1109/cec.2010.5585965

33. Petersen, B.K.: Deep symbolic regression: recovering mathematical expressions from data via policy gradients. arXiv abs/1912.04871 (2019). http://arxiv.org/abs/1912.04871

34. Pirkul, H.: A heuristic solution procedure for the multiconstraint zero-one knapsack problem. Nav. Res. Logist. **34**(2), 161–172 (1987)

35. Qian, F., Ding, R.: Simulated annealing for the 0/1 multidimensional knapsack problem. Numer. Math.-Engl. Ser. **16**(10201026), 1–7 (2007)

36. Sahoo, S., Lampert, C., Martius, G.: Learning equations for extrapolation and control. In: Dy, J., Krause, A. (eds.) Proceedings of the 35th International Conference on Machine Learning. Proceedings of Machine Learning Research, vol. 80, pp. 4442–4450. PMLR, 10–15 July 2018

37. Senju, S., Toyoda, Y.: An approach to linear programming with 0–1 variables. Manag. Sci. **15**(4), B-196-B-207 (1968). https://doi.org/10.1287/mnsc.15.4.b196

38. Sundar, S., Singh, A., Rossi, A.: An artificial bee colony algorithm for the 0–1 multidimensional knapsack problem. In: Ranka, S., et al. (eds.) IC3 2010. CCIS, vol. 94, pp. 141–151. Springer, Heidelberg (2010). https://doi.org/10.1007/978-3-642-14834-7_14

39. Udrescu, S.M., Tegmark, M.: AI Feynman: a physics-inspired method for symbolic regression. Sci. Adv. **6**(16), eaay2631 (2020). https://doi.org/10.1126/sciadv.aay2631

40. Van Lon, R.R., Holvoet, T., Vanden Berghe, G., Wenseleers, T., Branke, J.: Evolutionary synthesis of multi-agent systems for dynamic dial-a-ride problems. In: Proceedings of the 14th Annual Conference Companion on Genetic and Evolutionary Computation, GECCO 2012, pp. 331–336. ACM, New York (2012). https://doi.org/10.1145/2330784.2330832

41. Volgenant, A., Zoon, J.A.: An improved heuristic for multidimensional 0-1 knapsack problems. J. Oper. Res. Soc. **41**(10), 963–970 (1990). https://doi.org/10.2307/2583274. http://www.palgrave-journals.com/doifinder/10.1057/jors.1990.148

42. Özcan, E., Başaran, C.: A case study of memetic algorithms for constraint optimization. Soft. Comput. **13**(8–9), 871–882 (2008). https://doi.org/10.1007/s00500-008-0354-4

Algorithm Selection for the Team Orienteering Problem

Mustafa Mısır[1,2(✉)], Aldy Gunawan[3], and Pieter Vansteenwegen[4]

[1] Duke Kunshan University, Suzhou, China
mustafa.misir@dukekunshan.edu.cn
[2] Istinye University, Istanbul, Turkey
[3] Singapore Management University, Singapore, Singapore
aldygunawan@smu.edu.sg
[4] KU Leuven, Leuven, Belgium
pieter.vansteenwegen@kuleuven.be

Abstract. This work utilizes Algorithm Selection for solving the Team Orienteering Problem (TOP). The TOP is an NP-hard combinatorial optimization problem in the routing domain. This problem has been modelled with various extensions to address different real-world problems like tourist trip planning. The complexity of the problem motivated to devise new algorithms. However, none of the existing algorithms came with the best performance across all the widely used benchmark instances. This fact suggests that there is a performance gap to fill. This gap can be targeted by developing more new algorithms as attempted by many researchers before. An alternative strategy is performing Algorithm Selection that will automatically choose the most appropriate algorithm for a given problem instance. This study considers the existing algorithms for the Team Orienteering Problem as the candidate method set. For matching the best algorithm with each problem instance, the specific instance characteristics are used as the instance features. An algorithm Selection approach, namely ALORS, is used to conduct the selection mission. The computational analysis based on 157 instances showed that Algorithm Selection outperforms the state-of-the-art algorithms despite the simplicity of the Algorithm Selection setting. Further analysis illustrates the match between certain algorithms and certain instances. Additional analysis showed that the time budget significantly affects the algorithms' performance.

1 Introduction

Orienteering is essentially a type of sports, concerned with moving form a starting location to an end location while visiting a number of intermediate points. The goal is to maximize the total score which is collected through those visited points, within a given time limit. Orienteering is approached as an optimization problem in a general context beyond sports as the Orienteering Problem (OP) [12,15]. From this perspective, OP is modeled as a routing problem.

L. Pérez Cáceres and S. Verel (Eds.): EvoCOP 2022, LNCS 13222, pp. 33–45, 2022.
https://doi.org/10.1007/978-3-031-04148-8_3

Unlike pure routing attached to the Traveling Salesman Problem (TSP), the OP also includes the Knapsack Problem (KP) due to score maximization under a time budget. When the goal is to determine multiple paths, achieved by multiple people or vehicles, between the given starting and end points, the problem is denoted as the Team OP (TOP) [4]. Tourist trip planning is an example, real-world use case of the TOP [46]. The NP-hard nature of the TOP [36] has been attracting many researchers and practitioners to devise effective algorithmic solutions, mostly in the form of algorithms without optimality guarantee. Despite these development efforts, there is not a single algorithm outperforming all other algorithms for the TOP, as is the case for many other search and optimization problems [48].

One way to deal with this need for developing "the best" TOP algorithm is to specify the most suitable algorithm for each TOP instance. Algorithm Selection (AS) [22,40] is a systematic and automated way of achieving this task. AS is a high-level, meta-algorithmic strategy, traditionally achieving the selection tasks by deriving performance prediction models. These models basically map a set of features, $\mathcal{F}$, characterizing the problem instances, $\mathcal{I}$, to the algorithms, $\mathcal{A}$, performance, $\mathcal{P}_{|\mathcal{I}| \times |\mathcal{A}|}$. The resulting models are used to predict the performance of the candidate algorithms on a given problem instance. Regarding the candidate algorithms, Algorithm Portfolios (AP) [14] focus on having diverse and complementary algorithm sets to choose from. Such sets can also be used through scheduling [18] without AS. The schedules assign time periods to the candidate algorithms to run on one or more processing units. With sufficient computational resources, each AP member can run in parallel, returning the overall best solution(s). Going back to the AS models, the algorithm(s) with the best expected performance can be utilized to solve the target instance. Unlike this traditional use of AS, there have been AS methods with additional components such as pre-solvers [49,50]. While AS can be used to specify the best possible per-instance algorithm, there is not a single or an ultimate AS strategy. Considering that AS can be designed by utilizing different sub-components such as using a specific machine learning method as well as setting different parameter values for those accommodated sub-components, its automation was also studied, i.e. designing the best AS system based on the given AS tasks [25].

The present study applies AS to the TOP for the first time. The goal is to benefit from the existing algorithms by placing an AS layer on top of them. The algorithm set consists of 27 algorithms while the instance set used for both training and testing has 157 instances. For performing AS, the most obvious benchmark specifying elements are employed as $\mathcal{F}$. An existing AS system designed as an algorithm recommender system, i.e. ALORS [27], is used for the automated selection duty. Its problem instance and algorithm analysis capabilities besides its success in automatically choosing algorithms or their components is further shown on varying problems [28,29,31,32,39]. The experimental results revealed that AS is able to outperform all the constituent algorithms with almost no additional effort. Besides the pure selection results, a dis-/similarity analysis is reported both on the algorithm and instance space. All these analysis comes from the results of ALORS.

In the remainder of the paper, Sect. 2 further explains ALORS and how it is exactly used for this TOP setting. Section 3 reports the AS performance results while delivering the aforementioned analysis on the algorithms and the instances. The paper is summarized and the outcomes are briefly discussed together with listing the future research plans in Sect. 4.

2   Method

ALORS is a per-instance Algorithm Selection (AS) system based on recommender systems. ALORS specifically accommodates Collaborative Filtering (CF) [42]. Unlike the majority of the AS methods, CF allows ALORS to be able to work with incomplete performance data, $\mathcal{P}$. Yet, the present study works with the complete performance data. Considering the varying performance criteria, ALORS use $\mathcal{P}$ with the algorithms' ranks on each instance for general applicability. Another difference of ALORS is the way of building the performance prediction models. The mapping of instance features, $\mathcal{F}$, to $\mathcal{P}$ is indirectly achieved.

For this purpose, ALORS follows an intermediate step of feature to feature mapping. The initial, source features are those problem features, $\mathcal{F}$. The features to be mapped, i.e. latent (hidden) features $\mathcal{F}_h$, are automatically extracted from $\mathcal{P}$. As these features are directly driven from $\mathcal{P}$, they are expected to fairly represent the algorithms' performance. In that sense, ALORS takes the initial feature space to a new and more reliable feature space. This feature extraction process is handled by a well-known Matrix Factorization (MF) approach from linear algebra, i.e. Singular Value Decomposition (SVD) [13]. SVD is commonly used in CF or for dimensionality reduction as a data preprocessing step.

SVD applied to $\mathcal{P}_{|\mathcal{I}| \times |\mathcal{A}|}$ returns two main matrices besides a diagonal matrix of singular values, Σ. Referring to the components of $\mathcal{P}$, the initial matrix, U, consists of the latent features representing the problem instances. The other matrix, V, represents the candidate algorithms. They have been earlier used effectively on different problem domains [26,30]. Those matrices together approximates to $\mathcal{P}$ with the specified matrix ranks, $k \leq min(|\mathcal{I}|, |\mathcal{A}|)$, as follows. In our setting, r denotes the number of latent features. Since the singular values are sorted, from larger to smaller, earlier dimensions are more important than the latter dimensions, so the features. Thus, r can be picked as a small value while having strong approximation to $\mathcal{P}$ with possible noise elimination.

$$\mathcal{P} = U \Sigma V^T \approx U_k \Sigma V_k^T$$

When the mapping model is used for a new problem instance, it simply predicts a new row for U. This row is multiplied with Σ and V as denoted above. The outcome will be the rank predictions for all the candidate algorithms. Then, the algorithm with the best predicted rank is utilized for the target problem instance. The complete pseudo-code is provided in Algorithm 1.

Algorithm 1: ALORS: AS functionality [27]

Input

 Performance matrix in ranks $\mathcal{P} \in \mathbf{R}^{|\mathcal{I}| \times |\mathcal{A}|}$

 Initial representation $\mathcal{F} \in \mathbf{R}^{|\mathcal{I}| \times d}$ of the problem instances

 Features f of the target problem instance

AS Model Generation

 Build matrices U and V

 Build $\mathcal{E} = \{(f_i, U_i), i = 1 \dots |\mathcal{I}|\}$.

 Learn $\Phi : \mathbf{R}^d \mapsto \mathbf{R}^k$ from $\mathcal{E}$ (for a given matrix rank k)

AS Prediction

 Compute $U_f = \Phi(f)$

 Return $\underset{j=1\dots|\mathcal{A}|}{\arg\min} \ \langle U_f, V_j \rangle$

---

3 Computational Results

The Team Orienteering Problem (TOP) benchmarks are originated from [4]. The benchmark set consists of 387 instances. As the present performance data is collected from [16], only a subset of the instances are utilized, ignoring the ones where all the tested algorithms return the solutions of the same quality. This exclusion leads to 157 instances. The algorithm set has 27 approaches to be chosen as listed in Table 1.

The referenced study [16] reports the best solution delivered among 20 runs by each algorithm on each instance besides the average spent CPU time. The rank data is derived both based on the performance and the computational consumption. The spent time is used to break ties when two algorithms return the solution of the same quality. In other words, faster algorithms have better ranks than the slower algorithms with the same performance in terms of the solution quality. The features are simply the TOP benchmark specifications including n: the number of nodes, m: the number of vehicles and t_{max}: the maximum duration of each route.

For mapping, ALORS uses Random Forests (RF) [3]. SVD is performed with the rank of $k = 3$. The selection performance evaluation is achieved through 10-fold cross validation (10-cv). In other words, the TOP instance set of 157 instances is partitioned into 10 mutually exclusive, equally-sized[1] subsets. Each time, 9 subsets are used for training while the remaining subset involving unseen instances during training is used for testing.

Figure 1 illustrates the rank variations of each candidate TOP algorithm. While HALNS happens to be the overall best algorithm, MSA, AuLNS, PSOMA and MS-LS follow it. GLS, SiACO and DACO are the worst performing algorithms. Additionally, SkVNS and FPR are the least robust options as their rank performances significantly differ across the TOP instances. The detailed performance results can be found in the aforementioned study where the performance data is taken from [16].

---

[1] Almost equally-sized as 157 is not integer divisible by 10.

Table 1. 27 TOP algorithms

Algorithm	Variant
Tabu Search (TS) [9–11]	TMH [43], GTP [1], GTF [1]
Variable Neighborhood Search (VNS) [17,33]	FVNS [1], SVNS [1], SkVNS [45]
Ant Colony Optimization (ACO) [7]	SACO [19], DACO [19], RACO [19], SiACO [19],
Particle Swarm Optimization (PSO) [21]	PSOiA [6]
Local Search (LS)	GLS [45], MS-ILS [47], MS-LS [47]
Simulated Annealing (SA)	MSA [24]
GRASP [37] + Path Relinking (PR) [38]	FPR [41], SPR [41]
Large Neighborhood Search (LNS) [35]	AuLNS [23], LNS [34], HALNS [16]
Evolutionary Algorithms (EAs)	Memetic Algorithm (MA) [2], UHGS [47], UHGS-f [47]
Harmony Search (HS) [8]	SHHS [44], SHHS2 [44]
Hybrid	PSOMA [5]
Others	Pareto Mimic Algorithm (PMA) [20]

Table 2 reports the selection performance. Oracle, a.k.a. Virtual Best Solver (VBS), denotes the optimal performance by choosing the best algorithm for each instance. Single Best (SB) which is HALNS, represents overall best algorithm as one algorithm is used for all the instances. The results show that AS achieved by ALORS reaches to the overall average rank of 3.93 while the best algorithm out of 27 candidates come with the rank of 4.77. Take note that the lower the rank, the better the performance. Additionally, when the performance is evaluated with respect to each instance set, ALORS delivers the average rank of 4.18. The average rank of HALNS gets much worse, i.e. 4.77. These values indicate the clear advantage of ALORS. Being said that Oracle achieves the average rank of 1.12 and 1.10 for the respective evaluations. This means that there is still a room for improvement, likely by extending the instance feature space.

Figure 2 reveals the contribution of each feature on the AS prediction model of ALORS. RF used in ALORS, is a decision-tree based ensemble strategy providing importance on each single feature in terms of Gini importance/index. The corresponding importance evaluation shows that t_{max} happens to be the significantly most critical feature contributing to the AS recommendation model. Yet, although m and n are illustrated as substantially less important features than t_{max}, they still affect the algorithms' performance ranks. Being said that the use of t_{max} in the selection decisions may not be dominant if the benchmark instances are further diversified in terms of t_{max}.

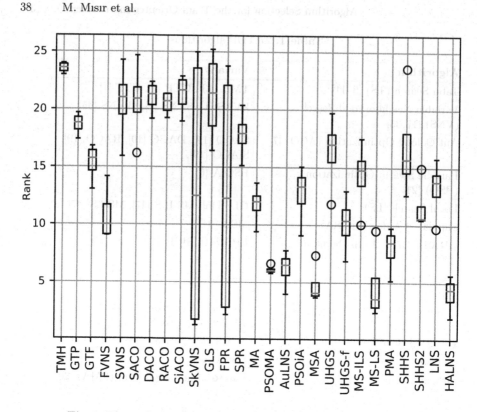

Fig. 1. The ranks of the TOP algorithms across all the instances

Table 2. The average ranks of each constituent algorithm besides ALORS where the best per-benchmark performances are in bold (#: the size of the instance set; SB: the Single Best solver; AVG: the average rank considering the average performance on each benchmark function; O-AVG: the overall average rank across all the instances)

Inst. Set (#)	Oracle	SB (HALNS)	ALORS
4 (54)	1.19	3.12	**3.02**
5 (45)	1.07	**5.18**	5.49
6 (15)	1.00	7.57	**5.23**
7 (43)	1.13	3.20	**2.99**
AVG	1.10	4.77	**4.18**
O-AVG	1.12	4.16	**3.93**

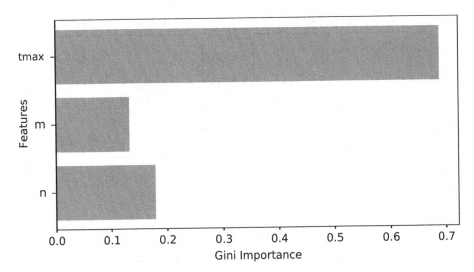

Fig. 2. Importance of the initial, hand-picked TOP instance features, using Gini Index/Importance

Further on the instance features, Fig. 3 visualizes the characterization of the instances through k-means clustering both considering those initial 3 basic features and 3 SVD originated latent features. The number of clusters, i.e. k, is determined by the best mean Silhouette score regarding the latent features. The features spaces are degraded into 2 via Multi-dimensional scaling (MDS). While the initial features yield rather clear instance distribution, the instance space looks more dispersed with the latent features. This is anticipated as the initial feature space is limited, only using 3 straightforward features. Instance by instance resemblance is depicted in Fig. 4.

Figure 5 visualizes the algorithms when they are hierarchically clustered, using the SVD ($k = 3$) driven features. For example, the leading single algorithm, i.e. HALNS, resembles to AuLNS and PMA. Essentially, in terms of performance HALNS and AuLNS similar as were discussed on Fig. 1. Unlike those comments, PMA is relatively poor compared both. Being said that the similarity comes from rank variation across the instances, not specifically the rank values. This aspect is reflected in Fig. 1 as the boxplot shapes are substantially similar.

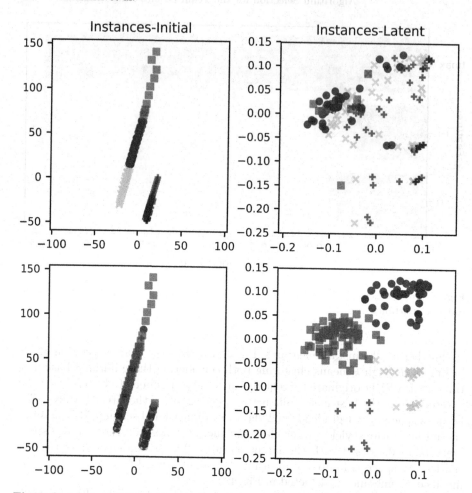

Fig. 3. Instances in 2-dimensional space using the initial and latent features (coloring is achieved by the initial features (top) and the latent features (bottom))

Besides these algorithms, PSOiA and LNS are determined as the most similar algorithms. Additionally, earlier mentioned least robust algorithms, i.e. SkVNS and FPR, are off the chart as they are cleary different than the rest, as shown in Fig. 1.

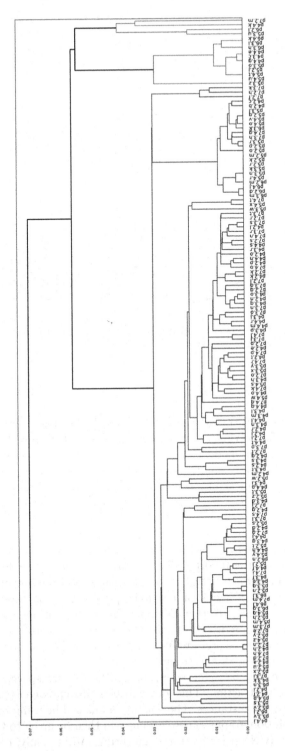

Fig. 4. Hierarchical clusters of instances using the latent features extracted from the performance data by SVD ($k = 3$)

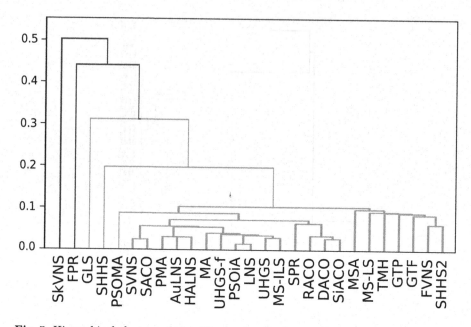

Fig. 5. Hierarchical clusters of algorithms using the latent features extracted from the performance data by SVD ($k = 3$)

4   Conclusion

Algorithm Selection (AS) is a meta-level approach allowing to benefit from multiple algorithms for solving a given target problem. Traditionally, AS delivers performance prediction models for a group of algorithms on an instance set, per-instance basis. This study applies AS to the Team Orienteering Problem (TOP), using an existing AS method named ALORS [27]. The task is to automatically determine the expectedly best algorithm among 27 candidate algorithms for any given TOP instance. Using the basic TOP benchmark specifications as the instance features, ALORS showed that all the constituent TOP algorithms are outperformed. Considering the diversity of the algorithms and the relatively large instance space, the reported dis-/similarity analysis provides a view both on the nature and characteristics of the algorithms and the instances.

This paper raises a series of research questions to be investigated. The reported research will be further extended by incorporating new features for strengthening the AS prediction quality, for example, whether nodes are clustered, nodes with higher scores are further from the start/end point, nodes with higher scores are clustered, and so on.

The next step will be achieved again on the feature space, yet targeting automated feature extraction. This idea will be realized by constructing images representing the TOP instances. The starting point will be using specific heuristics to deliver solution graphs. These graphs will then be given to Convolutional Neural Networks (CNNs) for feature engineering. Additionally, common graph based features will also be utilized. Finally, the computational results will be enriched by the other existing AS systems.

Acknowledgement. This study was supported by a Reintegration Grant project (119C013) of Scientific and Technological Research Council of Turkey (TUBITAK 2232).

References

1. Archetti, C., Hertz, A., Speranza, M.G.: Metaheuristics for the team orienteering problem. J. Heuristics **13**(1), 49–76 (2007). https://doi.org/10.1007/s10732-006-9004-0
2. Bouly, H., Dang, D.C., Moukrim, A.: A memetic algorithm for the team orienteering problem. 4OR **8**(1), 49–70 (2010). https://doi.org/10.1007/s10288-008-0094-4
3. Breiman, L.: Random forests. Mach. Learn. **45**(1), 5–32 (2001). https://doi.org/10.1023/A:1010933404324
4. Chao, I.M., Golden, B.L., Wasil, E.A.: The team orienteering problem. Eur. J. Oper. Res. **88**(3), 464–474 (1996)
5. Dang, D.-C., Guibadj, R.N., Moukrim, A.: A PSO-based memetic algorithm for the team orienteering problem. In: Di Chio, C., et al. (eds.) EvoApplications 2011. LNCS, vol. 6625, pp. 471–480. Springer, Heidelberg (2011). https://doi.org/10.1007/978-3-642-20520-0_48
6. Dang, D.C., Guibadj, R.N., Moukrim, A.: An effective PSO-inspired algorithm for the team orienteering problem. Eur. J. Oper. Res. **229**(2), 332–344 (2013)
7. Dorigo, M., Birattari, M., Stutzle, T.: Ant colony optimization. IEEE Comput. Intell. Mag. **1**(4), 28–39 (2006)
8. Geem, Z.W., Kim, J.H., Loganathan, G.V.: A new heuristic optimization algorithm: harmony search. SIMULATION **76**(2), 60–68 (2001)
9. Gendreau, M., Potvin, J.Y.: Tabu search. In: Burke, E.K., Kendall, G. (eds.) Search Methodologies, pp. 165–186. Springer, Boston (2005). https://doi.org/10.1007/0-387-28356-0_6
10. Glover, F.: Tabu search-part I. ORSA J. Comput. **1**(3), 190–206 (1989)
11. Glover, F.: Tabu search-part II. ORSA J. Comput. **2**(1), 4–32 (1990)
12. Golden, B.L., Levy, L., Vohra, R.: The orienteering problem. Nav. Res. Logist. (NRL) **34**(3), 307–318 (1987)
13. Golub, G.H., Reinsch, C.: Singular value decomposition and least squares solutions. Numer. Math. **14**(5), 403–420 (1970). https://doi.org/10.1007/BF02163027
14. Gomes, C., Selman, B.: Algorithm portfolio design: theory vs. practice. In: Proceedings of the 13th Conference on Uncertainty in Artificial Intelligence (UAI), Providence/Rhode Island, USA, 1–3 August 1997, pp. 190–197 (1997)
15. Gunawan, A., Lau, H.C., Vansteenwegen, P.: Orienteering problem: a survey of recent variants, solution approaches and applications. Eur. J. Oper. Res. **255**(2), 315–332 (2016)
16. Hammami, F., Rekik, M., Coelho, L.C.: A hybrid adaptive large neighborhood search heuristic for the team orienteering problem. Comput. Oper. Res. **123**, 105034 (2020)
17. Hansen, P., Mladenović, N., Todosijević, R., Hanafi, S.: Variable neighborhood search: basics and variants. EURO J. Comput. Optim. **5**(3), 423–454 (2016). https://doi.org/10.1007/s13675-016-0075-x

18. Kadioglu, S., Malitsky, Y., Sabharwal, A., Samulowitz, H., Sellmann, M.: Algorithm selection and scheduling. In: Lee, J. (ed.) CP 2011. LNCS, vol. 6876, pp. 454–469. Springer, Heidelberg (2011). https://doi.org/10.1007/978-3-642-23786-7_35

19. Ke, L., Archetti, C., Feng, Z.: Ants can solve the team orienteering problem. Comput. Ind. Eng. **54**(3), 648–665 (2008)

20. Ke, L., Zhai, L., Li, J., Chan, F.T.: Pareto mimic algorithm: an approach to the team orienteering problem. Omega **61**, 155–166 (2016)

21. Kennedy, J., Eberhart, R.: Particle swarm optimization. In: Proceedings of ICNN 1995-International Conference on Neural Networks, vol. 4, pp. 1942–1948. IEEE (1995)

22. Kerschke, P., Hoos, H.H., Neumann, F., Trautmann, H.: Automated algorithm selection: survey and perspectives. Evol. Comput. **27**(1), 3–45 (2019)

23. Kim, B.I., Li, H., Johnson, A.L.: An augmented large neighborhood search method for solving the team orienteering problem. Expert Syst. Appl. **40**(8), 3065–3072 (2013)

24. Lin, S.W.: Solving the team orienteering problem using effective multi-start simulated annealing. Appl. Soft Comput. **13**(2), 1064–1073 (2013)

25. Lindauer, M., Hoos, H.H., Hutter, F., Schaub, T.: AutoFolio: an automatically configured algorithm selector. J. Artif. Intelli. Res. **53**, 745–778 (2015)

26. Mısır, M.: Matrix factorization based benchmark set analysis: a case study on HyFlex. In: Shi, Y., et al. (eds.) SEAL 2017. LNCS, vol. 10593, pp. 184–195. Springer, Cham (2017). https://doi.org/10.1007/978-3-319-68759-9_16

27. Mısır, M., Sebag, M.: ALORS: an algorithm recommender system. Artif. Intell. **244**, 291–314 (2017)

28. Mısır, M.: Algorithm selection across selection hyper-heuristics. In: The Data Science for Optimization (DSO)@ IJCAI Workshop at the 29th International Joint Conference on Artificial Intelligence (IJCAI) (2021)

29. Mısır, M.: Algorithm selection on adaptive operator selection: a case study on genetic algorithms. In: Simos, D.E., Pardalos, P.M., Kotsireas, I.S. (eds.) LION 2021. LNCS, vol. 12931, pp. 237–251. Springer, Cham (2021). https://doi.org/10.1007/978-3-030-92121-7_20

30. Mısır, M.: Benchmark set reduction for cheap empirical algorithmic studies. In: IEEE Congress on Evolutionary Computation (CEC), pp. 871–877. IEEE (2021)

31. Misir, M.: Generalized automated energy function selection for protein structure prediction on 2D and 3D HP models. In: IEEE Symposium Series on Computational Intelligence (SSCI), pp. 1–6. IEEE (2021)

32. Mısır, M.: Selection-based per-instance heuristic generation for protein structure prediction of 2D HP model. In: IEEE Symposium Series on Computational Intelligence (SSCI), pp. 1–6. IEEE (2021)

33. Mladenovic, N., Hansen, P.: Variable neighborhood search. Comput. Oper. Res. **24**(11), 1097–1100 (1997)

34. Orlis, C., Bianchessi, N., Roberti, R., Dullaert, W.: The team orienteering problem with overlaps: an application in cash logistics. Transp. Sci. **54**(2), 470–487 (2020)

35. Pisinger, D., Ropke, S.: Large neighborhood search. In: Gendreau, M., Potvin, J.Y. (eds.) Handbook of Metaheuristics. ISOR, vol. 146, pp. 399–419. Springer, Boston (2010). https://doi.org/10.1007/978-1-4419-1665-5_13

36. Poggi, M., Viana, H., Uchoa, E.: The team orienteering problem: formulations and branch-cut and price. In: 10th Workshop on Algorithmic Approaches for Transportation Modelling, Optimization, and Systems (ATMOS 2010). Schloss Dagstuhl-Leibniz-Zentrum fuer Informatik (2010)

37. Resende, M.G., Ribeiro, C.: Greedy randomized adaptive search procedures (GRASP). AT&T Labs Research Technical Report **98**(1), 1–11 (1998)
38. Resendel, M.G., Ribeiro, C.C.: Grasp with path-relinking: recent advances and applications. In: Ibaraki, T., Nonobe, K., Yagiura, M. (eds.) Metaheuristics: Progress as Real Problem Solvers. ORCS, vol. 32, pp. 29–63. Springer, Boston (2005). https://doi.org/10.1007/0-387-25383-1_2
39. Ribeiro, J., Carmona, J., Mısır, M., Sebag, M.: A recommender system for process discovery. In: Sadiq, S., Soffer, P., Völzer, H. (eds.) BPM 2014. LNCS, vol. 8659, pp. 67–83. Springer, Cham (2014). https://doi.org/10.1007/978-3-319-10172-9_5
40. Rice, J.: The algorithm selection problem. Adv. Comput. **15**, 65–118 (1976)
41. Souffriau, W., Vansteenwegen, P., Berghe, G.V., Van Oudheusden, D.: A path relinking approach for the team orienteering problem. Comput. Oper. Res. **37**(11), 1853–1859 (2010)
42. Su, X., Khoshgoftaar, T.M.: A survey of collaborative filtering techniques. Adv. Artif. Intell. **2009**, 4 (2009)
43. Tang, H., Miller-Hooks, E.: A tabu search heuristic for the team orienteering problem. Comput. Oper. Res. **32**(6), 1379–1407 (2005)
44. Tsakirakis, E., Marinaki, M., Marinakis, Y., Matsatsinis, N.: A similarity hybrid harmony search algorithm for the team orienteering problem. Appl. Soft Comput. **80**, 776–796 (2019)
45. Vansteenwegen, P., Souffriau, W., Berghe, G.V., Van Oudheusden, D.: A guided local search metaheuristic for the team orienteering problem. Eur. J. Oper. Res. **196**(1), 118–127 (2009)
46. Vansteenwegen, P., Souffriau, W., Berghe, G.V., Van Oudheusden, D.: Iterated local search for the team orienteering problem with time windows. Comput. Oper. Res. **36**(12), 3281–3290 (2009)
47. Vidal, T., Maculan, N., Ochi, L.S., Vaz Penna, P.H.: Large neighborhoods with implicit customer selection for vehicle routing problems with profits. Transp. Sci. **50**(2), 720–734 (2016)
48. Wolpert, D., Macready, W.: No free lunch theorems for optimization. IEEE Trans. Evol. Comput. **1**, 67–82 (1997)
49. Xu, L., Hutter, F., Hoos, H., Leyton-Brown, K.: SATzilla: portfolio-based algorithm selection for SAT. J. Artif. Intell. Res. **32**(1), 565–606 (2008)
50. Xu, L., Hutter, F., Shen, J., Hoos, H., Leyton-Brown, K.: SATzilla 2012: improved algorithm selection based on cost-sensitive classification models. In: Proceedings of SAT Challenge 2012: Solver and Benchmark Descriptions, pp. 57–58 (2012)

Performance Evaluation of a Parallel Ant Colony Optimization for the Real-Time Train Routing Selection Problem in Large Instances

Bianca Pascariu[1]([✉]), Marcella Samà[1], Paola Pellegrini[2], Andrea D'Ariano[1], Joaquin Rodriguez[3], and Dario Pacciarelli[1]

[1] Department of Engineering, Roma Tre University,
Via della Vasca Navale 79, 00146 Rome, Italy
{bianca.pascariu,marcella.sama,andrea.dariano,
dario.pacciarelli}@uniroma3.it
[2] Univ. Lille Nord de France, Ifsttar, COSYS, LEOST, rue Élisée Reclus 20,
59666 Villeneuve d'Ascq, Lille, France
paola.pellegrini@univ-eiffel.fr
[3] Univ. Lille Nord de France, Ifsttar, COSYS, ESTAS, rue Élisée Reclus 20,
59666 Villeneuve d'Ascq, Lille, France
joaquin.rodriguez@univ-eiffel.fr

Abstract. The real-time Train Routing Selection Problem (rtTRSP) is the combinatorial optimization problem of selecting, for each train in a rail network, the best routing alternatives. Solving the rtTRSP aims to limit the search space of train rescheduling problems, highly affected by the number of routing variables. The rtTRSP is modelled as the minimum weight clique problem in an undirected k-partite graph. This problem is NP-hard and the problem size strongly affects the time required to compute optimal solutions. A sequential version of the Ant Colony Optimization (ACO) for the rtTRSP has been proposed in the literature. However, in large instances the algorithm struggles to find high-quality solutions. This paper proposes the performance evaluation of a parallel ACO for large rtTRSP instances. Specifically, we analyze the performance of the algorithm by standard parallel metrics. In addition, we test two parallel local search strategies, which consider different solution neighbourhoods. Computational experiments are performed on the practical case study of Lille Flandres station area in France. The results show a significant speed-up of the rtTRSP search space exploration and more promising diversification patterns. Both these improvements enhance the rtTRSP solutions.

Keywords: Ant Colony Optimization · Parallel computing · Rail Transportation · Scheduling and optimization of transportation systems

1 Introduction

During real-time operations, train timetables are highly perturbed by unexpected events. In these cases, detailed train rescheduling decisions have to be

L. Pérez Cáceres and S. Verel (Eds.): EvoCOP 2022, LNCS 13222, pp. 46–61, 2022.
https://doi.org/10.1007/978-3-031-04148-8_4

quickly taken by dispatchers. The train rescheduling problem consists in retiming, reordering and rerouting trains in such a way that the propagation of disturbances in the railway network is minimized. This problem is NP-hard [20]: solving the entire problem involves a high computational effort in large instances. However, the real-time nature of the problem requires very short computational times.

Among the methods in the literature to simplify the model and the solution process [3], in order to achieve high quality solutions in the short computation times, a possibility is to limit the number of train routing variables. These variables indicate whether a route is assigned to a train or not. The instance size of train rescheduling problems are highly affected by the number of routing variables [18], especially in complex railway stations: large infrastructures, high number of trains and available routes result in a huge number of constraints and variables. A key to come up with better solutions is to identify promising alternative routes in advance [5,18]. This problem is known in the literature as the real-time Train Routing Selection Problem (rtTRSP).

The rtTRSP is the combinatorial optimization problem of selecting, for each train in a rail network, the best routing alternatives. The rtTRSP is a rather neglected research topic. Recent works [21,26] model the problem as the minimum weight clique problem in an undirected k-partite graph. A clique is a classical concept in graph theory defined as a subset of adjacent vertices. Each partition of the graph is the set of routes corresponding to one train, with k number of trains. The vertex and edge weights represent the non-negative costs due to the route choices, in terms of conflict occurrence and scheduling decision impact. The rtTRSP optimal solution translates in finding the k-vertex clique of minimum weight (i.e., the best train routes combination), measured as the sum of weights of vertices and edges included in the clique. This problem is NP-hard [28], thus meta-heuristic algorithms are preferred since they are able to provide sub-optimal solutions within an acceptable time. On the contrary, exact algorithms fail to solve large instances within a reasonable time [21].

An Ant Colony Optimization (ACO) algorithm was recently proposed for the rtTRSP [26]. Compared to an exact algorithm for the same problem [21], ACO proves to converge to the global optimum in small instances. However, as the size of the instances increases, the algorithm struggles to find high-quality solutions.

This paper proposes the performance evaluation of a parallel ACO for large rtTRSP instances. We carefully analyze the parallel ACO-rtTRSP algorithm behaviour by standard metrics for parallel computing [22], such as speed, efficiency, and serial fraction. We also test two parallel local search strategies, which consider different solution neighbourhoods. Computational experiments are performed on the practical case study of Lille Flandres station area in France. The results show that the parallel ACO algorithm significantly speed-up the rtTRSP search space exploration. In addition, the two parallel local search strategies provide new exploration patterns which improve the result quality.

The rest of the paper is structured as follows: Sect. 2 gives a brief overview of related work; Sect. 3 presents the rtTRSP and its formal description; Sect. 4 describes the parallel ACO algorithm and the two local search strategies; Sect. 5 presents the computational analysis and the results, and Sect. 6 summarizes the conclusions and suggests where to focus future research on.

2   Related Work

Train rescheduling is a topic that has received significant research attention [3] since it belongs to the class of NP-hard problems [20]. Advanced techniques have been developed over time based on Mixed Integer Linear Programming (MILP) [23,25], the alternative graph [8,9,27], and the resource conflict graph [5]. Still, very large and difficult instances are unlikely to be solved to optimum in the short computational time required by the problem [25]. The alternative routes available for each train strongly affect the problem size and the required computation time: a partial routing flexibility is to be preferred to a complete routing flexibility [12,26].

Train routing is an NP-complete problem as soon as each train has three (or more) routing possibilities [18]. Recent approaches use subsets of all possible alternative routes, which are either based on guidelines set by the infrastructure managers [1,5], randomly selected [12], or chosen because considered the ones that will probably lead to the best quality solutions [26]. The problem of selecting optimized subset of alternative train routes has been formalized as the rtTRSP [26].

The rtTRSP is a rather neglected topic, recently modelled as the minimum weight clique problem in an undirected k-partite graph [21,26]. This problem is an NP-hard problem [28], closely related to the maximum clique problem and its generalizations. These are classical problems in graph theory, for which a wide variety of solution methods have been proposed [2,15]. The most popular are exact methods mainly based on branch-and-bound and variants of it [4,19,30]. However, these methods struggle to solve large instances in a short computational time. Effective heuristic [13,14,24] and meta-heuristic [10,16,29] algorithms have been proposed for these cases.

Recently, the rtTRSP was successfully solved by an ACO algorithm [26] inspired by one developed for the maximum clique problem [29]. This ACO-rtTRSP algorithm converges to the global optimum for small instances [21]. However, in large instances, the algorithm finds it difficult to effectively explore the solution space in real-time. In this paper, we investigate the opportunity offered by parallel programming [22,32] to improve the problem solution.

3   The rtTRSP

The rtTRSP is the combinatorial optimization problem of selecting the best subset of routes for each train in a network. The route selection is based on costs associated to scheduling decision impact, in terms of train conflicts. The rtTRSP solution is given as input to train rescheduling problem to reduce its search space and to provide better rescheduling solutions.

We consider a microscopic representation of the rail infrastructure, in which the *track-circuit* is the minimum infrastructure section that is able to detect the presence of a train within the fixed block signaling system. A *train route* is an ordered list of all track-circuits that a train passes through to reach the locations where the train has a scheduled stop, i.e., its *stopping points*. A set of train alternative routes share the same stopping points, while crossing different track-circuits. *Infrastructure utilization* is the time duration over which a track-circuit is allocated exclusively to a specific train, and blocked for the other ones [7].

During real-time operations, unexpected events may cause train delays (*primary delays*), which can lead to an infrastructure utilization overlap by two trains. The simultaneous utilization of the same track-circuit(s) is prevented by the signaling system, which stops one of the two concurrent trains. This train will inevitably suffer a delay, called *secondary delay*. Train rescheduling aims to minimize the total train secondary delays. When rerouting trains is contemplated, the rtTRSP estimates the secondary delay caused by the train route choice. The objective of the rtTRSP is thus to recommend high quality train route combinations which enforce the objective function of the train rescheduling problem.

3.1   Model Formulation

Let us consider a set T of k trains requiring to traverse a railway infrastructure within a certain time window. For each train $t \in T$, the set of all alternative route assignments is given. Let $G = (V, E)$ be an undirected k-partite graph, where V is the set of vertices and E is the set of edges. A vertex $v \in V$ in this graph represents an alternative train route. The vertices are grouped into k partitions such that for each train $t \in T$ we have an independent set $V_t \subset V$ of (all) alternative route assignments, i.e., $\cup_{t=1}^{k} V_t = V$ and $\cap_{t=1}^{k} V_t = \emptyset$. Two vertices $v_i, v_j \in V$ are connected by an edge $e_{ij} \in E$ if they belong to different trains and represent coherent routes. For a pair of trains, their routes are *coherent* in case no rolling stock re-utilization constraint exists between them. The latter constraints model turn-around, join, or split operations between two or more trains. If such constraint exists, the two routes are coherent when they satisfy the rolling stock constraint from an infrastructure point of view: the infrastructure section where the re-utilization takes place must be the same for all the trains involved in the constraint. The k-vertex cliques in this graph identify the set Γ of all feasible combinations of train routes. A clique is a complete (induced) sub-graph of G, such that every pair of the selected vertices is connected by an edge. The construction of a clique in G ensures the solution feasibility, i.e., one route has been selected for each train and each pair of train routes is coherent. In Fig. 1, the construction graph G for a 4-train example is presented. The graph is 4-partite: each train has a specific set of routes, indicated by the black dots. A 4-vertex clique is highlighted in bold.

Each vertex and edge in the graph has a non-negative weight, respectively, $u : V \rightarrow \mathbb{N}$ and $w : E \rightarrow \mathbb{N}$. Let us consider a train $t \in T$ and the set $V_t \subset V$

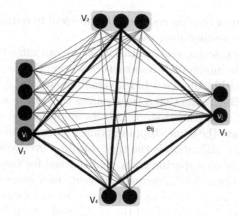

Fig. 1. Example of a construction graph $G = (V, E)$

of vertices associated to the alternative routes available for t. The weight u represents the delay due to the longer running time required to travel a route $v \in V_t$ compared to the timetable one $v_d \in V_t$. We calculate the weight u as $max(0, run_v - run_{v_d})$, where run_v indicates the minimum running time of v and run_{v_d} that of v_d. The weight w_{ij} assigned to edges, instead, is the potential delay when two coherent route assignments v_i and v_j, linked by e_{ij}, are jointly used. We define w_{ij} by assessing the minimum secondary train delay, based on the train ordering decision. The computation considers the expected entrance time in the network of the trains and their undisturbed running time until the conflict to be solved, if any. The total weight of a clique $c \in \Gamma$ is given by the sum $\sum_{v_i \in c} u_i + \sum_{e_{ij} \in c} w_{ij}$. Hence, the optimization problem consists of finding a clique with minimum weight.

4   Parallel ACO Meta-heuristic

ACO is a famous meta-heuristic which exploits the ant foraging behavior to solve hard combinatorial optimization problems [11]. Indirect communication between the ants enables them to find the shortest paths between their nest and food source. The algorithm adopts the concept of *pheromone trails* to keep track of the solution quality. Together with a *heuristic information*, a greedy measure of the expected quality of the solution to be found, the pheromone iteratively guides the solution space exploration from an incumbent solution to a hopefully better one.

A sequential implementation of ACO was applied to the rtTRSP [26], inspired by the ACO algorithm developed for the maximum clique problem [29]. A recent work [21] shows that the seminal ACO-rtTRSP [26] is able to find the global optimum, however when considering large instances, the algorithm does not converge in the short computation time available. To improve the performance on large instances and the quality solutions, we consider a parallel version of ACO-rtTRSP which relies on the multi-threading OpenMP *fork-join* model [6]. Multi-threading indicates that parallel computing is applied to threads which are part of a process. According to this parallel computing model, the investigated program starts as a single thread, called *master thread*, while at a designated point of its execution, this branches into a number of threads, creating a *parallel region*, and join later to resume the sequential execution on the master thread. Within the parallel region, each thread has both its *private memory* and a *shared memory*, whose access is available to all threads. During the sequential execution, the master thread uses the shared memory, while in the parallel region, the master gets its private memory.

The algorithm iteratively produces an rtTRSP solution until a certain time limit is reached or a zero-value objective function is found. Each iteration consists of the following three main steps:

- *ant walk* involves ant generation and initial solution construction, based on incremental selection of vertices;
- *daemon actions* includes the local search which improves the solutions from the ant walk;
- *pheromone update* modifies the pheromone trail, deposited by ants from the beginning of the search process on each edge of the construction graph G.

The computation of ant walk and pheromone update depends, respectively, on the total number of vertices and edges in the construction graph. In large problems, these steps significantly affect the total computation time. In comparison, the daemon actions is negligible in terms of time consumption but has a significant impact in terms of algorithmic performance.

The parallel computing is implemented in ant walk and pheromone update to speed-up the algorithm, since they are the most time-consuming parts. In addition, two local search strategies which exploit parallel computing improve diversification during the search space.

4.1 Speeding Up ACO: Parallel Ant Walk and Pheromone Update

In the ant walk step, the *nbAnts* ants of the colony run on multiple threads in parallel. Each ant of the colony is assigned to a thread and independently builds a solution on the construction graph G, i.e., a clique, within its private memory. A solution is built incrementally by adding one vertex v_i at a time among the available candidates v_h, i.e., the ones connected to each vertex in the partial solution c, chosen with a probability computed as the *random proportial rule* $\frac{\tau(v_i,c)^\alpha \eta(v_i,c)^\beta}{\sum_{v_h \in Candidates} \tau(v_h,c)^\alpha \eta(v_h,c)^\beta}$. Here, $\tau(v_i,c)$ indicates the sum of the pheromone trails associated to each e_{ij} such that $v_j \in c$, with α measuring the influence of pheromone trails in the random proportional rule. Furthermore, $\eta(v_i,c)$ indicates the sum of the heuristic information associated with each e_{ij} such that $v_j \in c$, computed as $\eta(v_{ij}) = 1/(1+w_{ij}+u_i)$, with β measuring the influence of heuristic information in the random proportional rule.

During the ant walk, each thread communicates with the master thread to select the best clique. The information about the best clique is placed in the shared memory and each thread accesses it in *read-write* mode. The concurrency of multiple threads can lead to a race condition, i.e., two or more threads can access and save their solution at the same time, in the same memory area, eventually causing a wrong algorithmic behaviour. To avoid this risk, we synchronize the threads such that the shared memory area allocated to the best clique is accessed one thread at a time. Other global information shared between the threads, as the construction graph and the pheromone matrix, is placed in the shared memory. Ants *read-only* this information such that no race condition can occur.

The best solution among those found by the entire ant colony is used in the daemon actions to move to better neighbouring solutions. The resulting best solution is then employed to update the pheromone trail. Here, an amount of $(1 - \rho)$ pheromones evaporates from all edges of G, with $0 < \rho < 1$ measuring the pheromone evaporation rate. At the same time, additional pheromone is deposited on the edges belonging to the best solution. An upper τ_{max} and lower τ_{min} bounds are imposed on the pheromone trails, following the MAX-MIN Ant System approach [31] which avoids the stagnation on few areas of the search space. Both evaporation and reinforcement actions are performed in parallel: pheromone matrix query is performed simultaneously by multiple threads. The pheromone matrix is placed in the shared memory. We manage the risk of overwriting the pheromone corresponding to parallel threads by allocating to each thread a specific area of the memory dedicated to the pheromone matrix.

4.2   Parallel Local Search Strategies

In the daemon actions, a local search is applied on the best solution(s) to move to better neighbouring one(s). Here, two alternative strategies, which employ parallel computing, are considered: (i) parallel cliques (*pCliques*), or (ii) parallel vertices (*pVertices*). The main local search process relies on two methods: *WorstVertex* and *BestVertex*, shown in Algorithm 1.

Given the construction graph G and a clique c, *WorstVertex* builds the ordered set *worstV* of vertices $v_i \in c$ such that the worst vertex v_w, i.e., the one with the highest weight calculated as $f(v_w) = max\{u_w + \sum_{i \in c} w_{wi}\}$ is the first. Regarding *BestVertex*, given a vertex $v_w \in worstV$, together with the graph G and the clique c, the function replaces vertex v_w with one of lower weight v_b, taken from the same partition of G, i.e., $v_b, v_w \in V_t$, in such a way that v_b forms a feasible and a lower weighted clique. The sequential local search strategy [26] consists of selecting the best clique at the end of the ant walk and replacing its worst vertex by *BestVertex*.

Let us consider n threads used to run the ACO-rtTRSP in its parallel form. *pCliques* strategy selects the n best cliques found by the ant colony during an iteration. Each clique is then assigned to a specific thread, which searches for the replacement of its worst vertex.

Algorithm 1: Local search

Data: Construction graph G, clique c

Result: Clique c of improved weight

1 WorstVertex (G, c)

2 **begin**

3   **forall** $v_i \in c$ **do**

4          $f(v_i) = u_i + \sum_{j \in c} w_{ij}$

5   sort in decreasing order of $f(v_i)$:

6      $worstV = \{v_i\}, \forall v_i \in c$

7 BestVertex $(G, c, v_w \in worstV)$

8 **begin**

9   **forall** $v_i \in V : v_w, v_i \in V_t, \exists e_{ij} \in G \; \forall v_j \neq v_w \in c$ **do**

10         $f(v_i) = u_i + \sum_{v_j \neq v_w \in c} w_{ij}$

11   **if** $f(v_i) < f(v_w)$ **then**

12             $v_b = v_i$

13   **if** $f(v_b) < f(v_w)$ **then**

14         $s \setminus \{v_w\} \cup \{v_b\}$

With *pVertices* strategy, the best clique c_b found during an iteration is selected. *WorstVertex* is then applied to the clique and the top n worst vertices in *worstV* are assigned each to a different thread, which independently performs *BestVertex*. To ensure no race condition, we allocate to each thread specific areas of *worstV*.

5   Computational Experiments

In this section, the performance of the parallel ACO-rtTRSP is carefully measured. The computational analysis is carried out on the practical test case of Lille Flandres station area in France. The infrastructure layout, shown in Fig. 2, is 12-km-long and composed of 299 track-circuits and 2409 routes. Lille Flandres station is a complex terminal station linked to national and international lines, with 17 platforms used by local, intercity, and high speed trains.

Starting from the one-day timetable the instances are obtained as follows. We perturb the initial timetable by applying a train delay between 5 and 15 min to the entrance time of 20% of trains. The trains to be delayed and the corresponding delay are randomly selected. For each perturbed timetable, we select ten instances corresponding to a 60-min time window of traffic flow. We select this instances by starting from ten different time instants, randomly drawn during the peak-hour time periods 7:30–9:00 and 18:30–20:00. We generate a total of 40 instances. In each instance, there are on average 39 trains per hour, and 170 routing alternatives per train. The corresponding construction graphs have an average of 6,368 vertices and 21,365,433 edges in total.

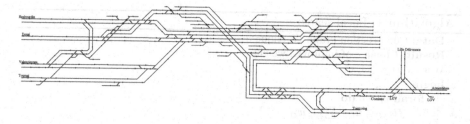

Fig. 2. Lille Flandres station area

Since the rtTRSP is designed to be solved in real-time, we give to the ACO-rtTRSP algorithm a maximum computation time of 30 s to stay compatible with train operations. The computational analysis is conducted in two parts. We first assess the algorithm speed-up in different setting configurations. For these experiments, we consider the sequential local search in Sect. 4. We previously assessed that the parallel local search strategies consume a comparable computation time to the sequential local search. In fact, the parallel strategies do not contribute to the algorithm speed-up, but to new search space exploration patterns. In the second part of the analysis, we evaluate the impact of these new exploration patterns on the solution quality.

ACO depends on the values of the user-defined parameters: α, the pheromone factor weight; β, the heuristic factor weight; ρ, the pheromone evaporation rate; and $nAnts$ the number of ants in the colony; τ_{max} an τ_{min}, the maximum and minimum pheromone bounds. Unless differently specified, we use the ACO-specific parameter setting: $nAnts = 200$, $\alpha = 2$, $\beta = 10$, $\rho = 0.09$, $\tau_{min} = 0.003$, and $\tau_{max} = 4$. In a preliminary analysis, we assessed different settings of these parameters. They do not appear to impact the conclusions that we can draw on the performance achieved when using a different number of threads or different type of local search strategy.

We perform the experiments on an Intel Xeon 22 core 2.2 GHz processor with 1.5 TB RAM, under Windows distribution, using C++ to implement the algorithm. The machine provides a maximum of 44 threads.

5.1   Results

We compare the profiling of the sequential (single thread) and parallel (42 threads) algorithm in Fig. 3. The graph represents the time spent by the three main steps - ant walk, daemon actions and pheromone update - during one iteration of the algorithm. The number of vertices in G is used to gauge the problem size. The left and right vertical axes report the computation time when using one and 42 threads, respectively.

The single thread profiler data in Fig. 3 show that the daemon action is negligible in terms of time. While ant walk and pheromone update are the most time consuming steps. These two steps require a comparable amount of time when the problem has smaller number of vertices in G. As the vertices increases,

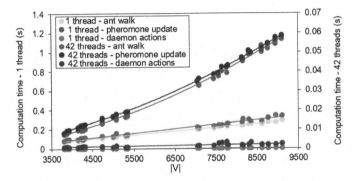

Fig. 3. Profiler data on time consumption by the different ACO algorithm steps.

the number of edges grows exponentially, raising the computation time required for the pheromone update compared to the ant walk: in the largest instances, the time of an iteration is consumed around 70% by pheromone update and 30% by ant walk. The parallel profiler data in Fig. 3 show the same trend of the sequential ones. However, the time required by each algorithm step is significantly reduced in parallel, as shown by the right axis scale. We remark that this time reduction is independent from the problem size.

Figure 4 reports the algorithm parallel performance when scaling the number of threads. We analyze the overall algorithm speed-up and efficiency in Fig. 4a, while we focus on the ant walk and pheromone update speed-up in Fig. 4b. Furthermore, we study the impact of different ant colony sizes on the ant walk performance, in terms of speed-up and serial fraction, in Fig. 4c and 4d. For the latter analysis, we run the algorithm with 100, 200, and 400 ants. The data represented in Fig. 4 are accompanied by the sample standard deviation.

We use the *speed-up*, the *efficiency* and the *serial fraction* metrics to evaluate the parallel performance. These are the most common metrics used by the research community [17,22]. The *speed-up* evaluates the extent at which the parallel algorithm is faster than the corresponding sequential algorithm. Since the algorithm always runs for a total computation time of 30 s, we consider the average time to perform one iteration. The speed-up is computed as Q_1/Q_n, where Q_1 is the serial execution time of one iteration, and Q_n the parallel execution time with n threads. By definition, the speed-up is *linear* when its value is equal to n. The *efficiency* is the normalized value of the speed-up and is defined as the ratio between the speed-up and the number of threads. The linear speed-up corresponds to an efficiency of one. The *serial fraction* is obtained experimentally using the Karp-Flatt metric [17] $(1/s - 1/n)/(1 - 1/n)$. In the formula, s is the speed-up when n threads are used. This metric takes into account the parallel *overhead* and helps to detect the sources of inefficiency with increasing n. The overhead is any combination of excessive or indirect computation time, memory, bandwidth, or other resources that are required to perform a specific task. The serial fraction would remain constant in an ideal system, while compu-

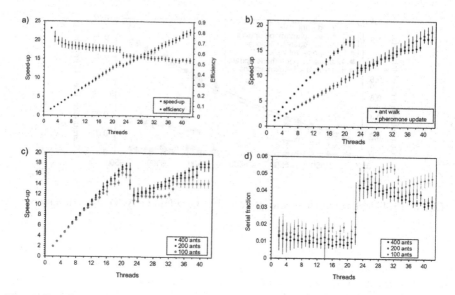

Fig. 4. Parallel performances with increasing number of threads: (a) algorithm speed-up and efficiency, (b) ant walk and pheromone update speed-up, (c) ant walk speed-up with 100, 200 and 400 ants or (d) ant walk serial fraction with 100, 200 and 400 ants.

tation time, speed-up, and efficiency vary with n. Irregular change in the serial fraction points to load-balancing issues, e.g., some threads take longer than others. Nearly constant serial fraction values, for all threads, indicates that the loss of efficiency is due to the limited parallelism of the program. While increasing values are usually caused by an overhead increase. A decrease of the serial fraction can also be encountered since adding threads can add cache and memory bandwidth.

The results in Fig. 4a show that the algorithm has a good *scalability*: the speed-up steadily grows with the number of threads. However, the speed-up results sub-linear and the efficiency decrease with the number of threads due to an increase in algorithmic or architectural overhead. In particular, we record a performance disruption with 22 threads.

The algorithm performance is better explained when considering the separate behaviour of ant walk and pheromone update in Fig. 4b. Here, we can see that the pheromone update has a uniformly sub-linear speed-up with a trend line slope of 0.5. The pheromone matrix has in general a huge size, equal to the number of edges in the construction graph. We then assume that the performance in accessing the pheromone matrix is deteriorated by an overhead in memory management or synchronization between threads. Moreover, Fig. 4b highlights that the performance loss around 22 threads is mainly caused by ant walk. Up to this point, ant walk is able to achieve an almost linear speed-up. After that, we can not tell exactly what this overhead is due to: machine configuration, memory management, communications or synchronization between threads.

The results in Fig. 4c points out a clear trend in ant walk parallel performance: regardless the number of ants, around 22 threads there is a loss of efficiency. Given the big increase of serial fraction in Fig. 4d in correspondence of this point, it is evident that the cause is a communication bottleneck. This is most probably due to the number of physical core of the machine used for the experiments.

Table 1. Statistics results on the objective function with increasing number of threads.

		Threads							
		1	6	12	18	24	30	36	42
15 s	Avg. obj. value	1910	1586	1222	1132	1118	1104	1042	996
	Avg. iterations	31	99	186	272	313	375	438	496
Δ_{0-15}	μ	1054	1139	1484	1589	1603	1604	1642	1688
	p-value	$3 \cdot 10^{-7}$	$4 \cdot 10^{-8}$	$4 \cdot 10^{-8}$	$4 \cdot 10^{-8}$	$4 \cdot 10^{-8}$	$4 \cdot 10^{-8}$	$4 \cdot 10^{-8}$	$4 \cdot 10^{-8}$
	LCI	588	833	1153	1222	1236	1253	1254	1274
	UCI	1546	1682	1946	1994	1994	1994	2106	2186
30 s	Avg. obj. value	1678	1215	1107	989	962	944	938	937
	Avg. iterations	61	198	376	546	626	750	876	996
Δ_{15-30}	μ	385	283	97	74	71	86	65	27
	p-value	$4 \cdot 10^{-5}$	$2 \cdot 10^{-6}$	$8 \cdot 10^{-7}$	$1 \cdot 10^{-6}$	$3 \cdot 10^{-6}$	$2 \cdot 10^{-6}$	$1 \cdot 10^{-6}$	$1 \cdot 10^{-6}$
	LCI	258	133	39	34	27	21	14	7
	UCI	561	617	127	213	163	179	157	83
$\Delta_{n-n'}$	μ	-	433	87	42	9	8	6	3
	p-value	-	$1 \cdot 10^{-6}$	$8 \cdot 10^{-7}$	$9 \cdot 10^{-6}$	$7 \cdot 10^{-4}$	$6 \cdot 10^{-5}$	$2 \cdot 10^{-4}$	$5 \cdot 10^{-3}$
	LCI	-	268	30	16	4	3	3	1
	UCI	-	689	118	214	16	25	94	4

In Table 1 we report the statistics on the evolution of the objective function when different number of threads are used. We represent on the columns the number of threads considered, while we record on the rows the best objective function, the number of iterations, and the objective function improvement (Δ_{0-15}, Δ_{15-30}) after 15 and 30 s of computation. Next, we measure the objective function improvement achieved from the gradual addition of threads ($\Delta_{n-n'}$). To this purpose, we compare for each number of threads (t) the result obtained in 30 s, with that of the previous column (n'). The improvements are obtained by performing the Wilcoxon signed rank test on the objective function difference. We use the test to determine whether the pseudo-median of the sample μ is equal to zero (null hypothesis) with a Confidence Interval (CI) of 95%. When this hypothesis is true, no significant improvement is recorded. Positive values of μ and of the lower (LCI) and upper (UCI) bounds of the CI means that the sample objective function improves significantly, vice versa it does not with negative values.

Looking at the results in Table 1, we can distinguish two phases of the algorithm behavior. In the first phase (Δ_{0-15}), diversification actions lead to a rapid improvement of the best-know solution, reflected by the high values of μ. In particular, as the number of threads grows, up to 18 threads, more significant improvements in the objective values are obtained, indicating that the algorithm parallelization improves the diversification actions. In the second phase (Δ_{15-30}), a rather slow improvement is achieved, except for the threads 1 and 8 which still present a remarkable improvement. Given the small number of iterations obtained with these threads, we can assume that diversification actions occur mainly within 300 iterations or so. The very small p-values, and the LCI and UCI trends similar to that of μ confirm the significance of the results.

The results related to $\Delta_{n-n'}$ in Table 1 show that adding threads brings a remarkable initial improvement of the objective function. The benefits of adding more threads decreases after 18 threads due to the speed-up slowdown. This leads to minor improvements of the objective function. In practice, the performance is satisfactory at about the same level, for any number of threads greater than 18. Or alternatively, it is possible to use 42 threads and 15 s of computation. While when looking for the optimal solution, it is however better to use as many threads as available and 30 s of computation, since the Wilcoxon test returns always a significant improvement of the objective function, with positive μ, LCI and UCI, and p-values smaller than the significance level alpha $= 0.05$.

Table 2. Local search performance comparison.

Local search	Objective value (s)	#Iter. (\30 s)	#Improv. (\#Iter.)	Improv. value (s)	μ	Δ obj. value p-value	(LCI, UCI)
None	1007	1012	0	0	-	-	-
Sequential	937	996	271	328	27	0.02	(3, 64)
pCliques	901	977	494	259	34	0.002	(11, 89)
pVertices	822	956	796	200	30	0.01	(3, 73)

We now focus the computational analysis on the diversification action carried out by the local research. Specifically, we compare the impact of the different local search strategies in Sect. 4.2 on the ACO-rtTRSP solution. We report the results of the comparison in Table 2. We run ACO with 42 threads, by alternately including in the daemon actions the local search strategy indicated in the first column of the table. We evaluate the average over all instances, of the best objective value and the number of iterations in the 30 s of computation, respectively, in the second and third column of the table. We remind that, at each iteration, ACO provides a solution that the local search seeks to improve. In the fourth column, we report the average number of times the local search improves the ACO solution, over the total number of iterations in each run of the algorithm. For these solutions, the next column of the table reports the average improvement, i.e., the difference between the ACO and the local search solution values. The reminder columns report the results of the Wilcoxon signed rank test on the objective function difference with a 95% CI. Such improvement is performed between each local search strategy and the one at the previous row.

From the results in Table 2, we can first observe that including a local search within the ACO algorithm typically generates significant improvements to the objective function value. The worst performance is achieved when no local search is considered. The parallelization, introduced in pCliques and pVertices, steadly improves the solution quality. The neighborhoods of the two parallel strategies are more promising than the sequential local search. The gradual decrease of the iterations in Table 2 shows that pCliques and pVertices require slightly higher computation time than the sequential local search. However, this is negligible for the solution quality. The improved cliques obtained by pVertices guide the solution process towards better parts of the search space, leading to the best objective values. pVertices improves the highest number of solutions: it most likely has a larger neighborhood than pCliques which allows a wider exploration of the search space. pVertices explores simultaneously all the graph partitions, so its neighborhood size is approximately $|V|(k-1)$. While when considering multiple cliques by pCliques the neighborhood may be smaller: the worst vertex of different cliques may belong to the same partition, and the cliques may differ by a small number of vertices.

6   Conclusions and Future Research

In this paper, we evaluated the performance of a parallel algorithm for large rtTRSP instances. The rtTRSP is modelled as the minimum weight clique problem in an undirected k-partite graph. This problem is NP-hard and its complexity is highly increased by the size of the instances and the short available computation time. We solved the problem by a parallel ACO algorithm to speed-up the search space exploration. In addition, we assessed two parallel local search strategies, to improve diversification during the solution search. We analyzed the parallel behaviour of the algorithm in terms of speed-up, efficiency and serial factor. The methodology and performance evaluation are applied to the practical rtTRSP problem, but they are equally applicable to all the problems modeled as the minimum weight clique problem in an indirect k-partite graph. Computational experiments were perform on the practical case study of Lille Flandres station area, in France.

The results show a significant improvement of the search space exploration. The algorithm has a good scalability in terms of parallel performance: the speed-up steadily grows with the number of threads, reaching a maximum value of 22 with 42 threads. This speed-up improves the objective function by 44% compared to the serial algorithm. However, the efficiency of the parallel implementation is compromised by the communication overhead when more threads than the physical cores are used. To overcome this shortcoming, future work may be focus on reducing the communication between the ants in the colony or to implement asynchronous communication. Moreover, the two parallel local search strategies provide wider neighborhoods, enlarging the search space exploration and improving the solution quality. Future research can be dedicated to apply our methodology to other rail infrastructures, with different characteristics from those presented here. Furthermore, other exact or meta-heuristic methods for the problem could be designed.

Acknowledgements. This work was supported by the "SPECIES Scholarships" with a three-month mobility grant for Ph.D. candidates.

References

1. Bettinelli, A., Santini, A., Vigo, D.: A real-time conflict solution algorithm for the train rescheduling problem. Transp. Res. Part B **106**, 237–265 (2017)
2. Bomze, I.M., Budinich, M., Pardalos, P.M., Pelillo, M.: The maximum clique problem. In: Du, D.Z., Pardalos, P.M. (eds.) Handbook of Combinatorial Optimization, pp. 1–74. Springer, Boston (1999). https://doi.org/10.1007/978-1-4757-3023-4_1
3. Cacchiani, V., et al.: An overview of recovery models and algorithms for real-time railway rescheduling. Transp. Res. Part B **63**, 15–37 (2014)
4. Cai, S., Lin, J.: Fast solving maximum weight clique problem in massive graphs. In: IJCAI, pp. 568–574 (2016)
5. Caimi, G., Chudak, F., Fuchsberger, M., Laumanns, M., Zenklusen, R.: A new resource-constrained multicommodity flow model for conflict-free train routing and scheduling. Transp. Sci. **45**(2), 212–227 (2011)
6. Chapman, B., Jost, G., Van Der Pas, R.: Using OpenMP: Portable Shared Memory Parallel Programming, vol. 10. MIT Press, Cambridge (2008)
7. Corman, F., D'Ariano, A., Pacciarelli, D., Pranzo, M.: Evaluation of green wave policy in real-time railway traffic management. Transp. Res. Part C **17**(6), 607–616 (2009)
8. Corman, F., D'Ariano, A., Pacciarelli, D., Pranzo, M.: A tabu search algorithm for rerouting trains during rail operations. Transp. Res. Part B **44**(1), 175–192 (2010)
9. D'Ariano, A., Pacciarelli, D., Pranzo, M.: A branch and bound algorithm for scheduling trains in a railway network. Eur. J. Oper. Res. **183**(2), 643–657 (2007)
10. Dang, D.C., Moukrim, A.: Subgraph extraction and metaheuristics for the maximum clique problem. J. Heuristics **18**(5), 767–794 (2012). https://doi.org/10.1007/s10732-012-9207-5
11. Dorigo, M., Stützle, T.: Ant Colony Optimization. MIT Press, Cambridge (2004)
12. Fischetti, M., Monaci, M.: Using a general-purpose mixed-integer linear programming solver for the practical solution of real-time train rescheduling. Eur. J. Oper. Res. **263**(1), 258–264 (2017)
13. Gimadi, E.K., Kel'manov, A.V., Pyatkin, A.V., Khachai, M.Y.: Efficient algorithms with performance guarantees for some problems of finding several cliques in a complete undirected weighted graph. Proc. Steklov Inst. Math. **289**(1), 88–101 (2015). https://doi.org/10.1134/S0081543815050089
14. He, G., Liu, J., Zhao, C.: Approximation algorithms for some graph partitioning problems. In: Graph Algorithms and Applications, vol. 2, pp. 21–31 (2004)

15. Hosseinian, S., Fontes, D.B.M.M., Butenko, S., Nardelli, M.B., Fornari, M., Curtarolo, S.: The maximum edge weight clique problem: formulations and solution approaches. In: Butenko, S., Pardalos, P.M., Shylo, V. (eds.) Optimization Methods and Applications. SOIA, vol. 130, pp. 217–237. Springer, Cham (2017). https://doi.org/10.1007/978-3-319-68640-0_10

16. Hudry, O.: Application of the descent with mutations metaheuristic to a clique partitioning problem. RAIRO-Oper. Res. **53**(3), 1083–1095 (2019)

17. Karp, A.H., Flatt, H.P.: Measuring parallel processor performance. Commun. ACM **33**(5), 539–543 (1990)

18. Kroon, L.G., Romeijn, H.E., Zwaneveld, P.J.: Routing trains through railway stations: complexity issues. Eur. J. Oper. Res. **98**(3), 485–498 (1997)

19. Kumlander, D.: A new exact algorithm for the maximum-weight clique problem based on a heuristic vertex-coloring and a backtrack search. In Proceedings of 5th International Conference on Modelling, Computation and Optimization in Information Systems and Management Sciences, pp. 202–208 (2004)

20. Mascis, A., Pacciarelli, D.: Job shop scheduling with blocking and no-wait constraints. Eur. J. Oper. Res. **143**(3), 498–517 (2002)

21. Pascariu, B., Samà, M., Pellegrini, P., D'Ariano, A., Pacciarelli, D., Rodriguez, J.: Train routing selection problem: ant colony optimization versus integer linear programming. IFAC-PapersOnLine **54**(2), 167–172 (2021)

22. Pedemonte, M., Nesmachnow, S., Cancela, H.: A survey on parallel ant colony optimization. Appl. Soft Comput. **11**(8), 5181–5197 (2011)

23. Pellegrini, P., Marlière, G., Rodriguez, J.: Optimal train routing and scheduling for managing traffic perturbations in complex junctions. Transp. Res. Part B **59**(1), 58–80 (2014)

24. Pullan, W.: Approximating the maximum vertex/edge weighted clique using local search. J. Heuristics **14**(2), 117–134 (2008). https://doi.org/10.1007/s10732-007-9026-2

25. Pellegrini, P., Presenti, R., Rodriguez, J.: Efficient train re-routing and rescheduling: valid inequalities and reformulation of RECIFE-MILP. Transp. Res. Part B **120**(1), 33–48 (2019)

26. Samà, M., Pellegrini, P., D'Ariano, A., Rodriguez, J., Pacciarelli, D.: Ant colony optimization for the real-time train routing selection problem. Transp. Res. Part B **85**(1), 89–108 (2016)

27. Samà, M., D'Ariano, A., Corman, F., Pacciarelli, D.: A variable neighbourhood search for fast train scheduling and routing during disturbed railway traffic situations. Comput. Oper. Res. **78**, 480–499 (2017)

28. Solano, G., Blin, G., Raffinot, M., Clemente, J.B., Caro, J.: On the approximability of the minimum weight t-partite clique problem. J. Graph Algorithms Appl. **24**(3), 171–190 (2020)

29. Solnon, C., Bridge, D.: An ant colony optimization meta-heuristic for subset selection problems. In: System Engineering Using Particle Swarm Optimization, pp. 7–29. Nova Science Publisher (2006)

30. Sørensen, M.M.: New facets and a branch-and-cut algorithm for the weighted clique problem. Eur. J. Oper. Res. **154**(1), 57–70 (2004)

31. Stützle, T., Hoos, H.H.: MAX-MIN ant system. Future Gener. Comput. Syst. **16**(8), 889–914 (2000)

32. Wu, Q., Spiryagin, M., Cole, C., McSweeney, T.: Parallel computing in railway research. Int. J. Rail Transp. **8**(2), 111–134 (2020)

Deep Infeasibility Exploration Method for Vehicle Routing Problems

Piotr Beling[1], Piotr Cybula[1,4(✉)], Andrzej Jaszkiewicz[2],
Przemysław Pełka[3,4], Marek Rogalski[1], and Piotr Sielski[1]

[1] Faculty of Mathematics and Computer Science, University of Lodz,
Banacha 22, 90-238 Lodz, Poland
[2] Faculty of Computing, Poznan University of Technology,
Piotrowo 3, 60-965 Poznan, Poland
[3] Faculty of Electrical, Electronic, Computer and Control Engineering,
Lodz University of Technology, Stefanowskiego 18/22, 90-924 Lodz, Poland
[4] Emapa S.A., Ciolka 12, 01-402 Warsaw, Poland
piotr.cybula@wmii.uni.lodz.pl

Abstract. We describe a new method for the vehicle routing problems with constraints. Instead of trying to improve the typical metaheuristics used to efficiently solve vehicle routing problems, like large neighborhood search, iterated local search, or evolutionary algorithms, we allow them to explore the deeply infeasible regions of the search space in a controlled way. The key idea is to find solutions better in terms of the objective function even at the cost of violation of constraints, and then try to restore feasibility of the obtained solutions at a minimum cost. Furthermore, in order to preserve the best feasible solutions, we maintain two diversified pools of solutions, the main pool and the temporary working pool. The main pool stores the best diversified (almost) feasible solutions, while the working pool is used to generate new solutions and is periodically refilled with disturbed solutions from the main pool. We demonstrate our method on the vehicle routing problems, with variants respecting time, vehicle capacity and fleet limitation constraints. Our method provided a large number of new best-known solutions on well-known benchmark datasets. Although our method is designed for the family of vehicle routing problems, its concept is fairly general and it could potentially be applied to other NP-hard problems with constraints.

Keywords: Metaheuristics · Combinatorial optimization ·
Transportation · Vehicle routing problem

1 Introduction

The family of vehicle routing problems (VRP) comprises NP-hard combinatorial optimization problems with high practical importance in transportation and

This work has been partially supported by the Polish National Centre for Research and Development (projects POIR.01.01.01-00-0222/16, POIR.01.01.01-00-0012/19).

logistics [3, 24]. Members of this family may involve time windows, heterogeneous fleet, pickup and deliveries, and multiple depots, while solutions to these problems need to meet vehicle capacity, time windows, precedence, and fleet limit constraints.

Typical metaheuristics used to efficiently solve vehicle routing problems include large neighborhood search [19], iterated local search [25], evolutionary algorithms [16, 26], and combinations of these methods. Such methods have, however, a tendency to get stuck in deep local optima after relatively short time, see e.g. [19]. Escaping from a deep local optimum may be very difficult since other very good local optima may be significantly different. For example, our tests performed on 400-customer [8] benchmark instances revealed that solutions very close to the best-known solution in terms of the objective function value (worse by less than 0.05%) may differ from the best solution by more than 100 edges (≈23%). On the one hand, such solutions are very similar since they have a majority of edges in common. On the other hand, obtaining one of these solutions from another via a series of small steps may be very difficult, since the intermediate solutions can be either much worse or highly infeasible, with one or more constraints highly violated.

In order to overcome this problem we propose to solve vehicle routing problems with a new *Deep Infeasibility Exploration Method* (*DIEM*). The main idea of our method is to allow typical metaheuristics used for VRP, like large neighborhood search, iterated local search, or evolutionary algorithms, to explore the deeply infeasible regions of the search space in a controlled way.

Our approach is motivated by the following observations:

- It may be beneficial to explore infeasible regions of the search space in order to obtain new promising solutions, i.e. solutions with good values of the objective function and having a structure that allows for a relatively easy restoration of the feasibility.
- In order to improve the solution which is currently the best on the objective function, a new solution (even infeasible) with a better value of the objective function needs to be found.

However, as the number of working solutions that can be simultaneously maintained by an algorithm is limited, there is a trade-off between accepting infeasible solutions and conserving the good feasible solutions. Thus, we propose to use two pools with proper diversity of solutions stored in each. The *working pool* is used by the underlying metaheuristics to generate new good (even highly infeasible) solutions. The *main pool* is used to store diverse good solutions lying near the border between feasible and infeasible regions. The main pool is updated with new solutions from the working pool if they are (almost) feasible. Whenever the working pool does not contain new promising solutions, it is emptied and refilled with disturbed solutions from the main pool.

Our method involves two interlacing phases:

1. *Improvement phase*, with the goal of generating new high quality solutions, even at the cost of infeasibility, starting with solutions from the main pool.

2. *Feasibility restoration phase*, with the goal of restoring feasibility of high quality solutions with a minimum deterioration of their quality, starting with solutions from the working pool.

In both phases, a penalty function approach is used. The penalty weights, however, differ significantly in each type of phase. In the improvement phase, the penalty weights are relatively low. They are adaptively set to obtain solutions better in terms of the objective function (without penalty terms). In the feasibility restoration phase, the penalty weights are higher, increasing the chance of obtaining feasible solutions.

However, it may be hard to escape the current region of the search space when starting from previously generated local optima, even with modified penalty weights. Thus, before starting a new optimization phase (of any type), we significantly disturb all solutions taken either from the working pool or the main pool. This way we prevent stagnation and allow penalty coefficients adjustment. Similar approach is used in other high quality methods for VRP, e.g. [2].

DIEM was implemented in a solver for a wide range of vehicle routing problems. Our method provided 114 new best-known solutions on [8] benchmark instances for the Vehicle Routing Problem with Time Windows (VRPTW), 130 new best-known solutions on [13] benchmark instances for the Pickup and Delivery Problem with Time Windows (PDPTW) and 39 new best-known solutions on benchmark instances for the Team Orienteering Problem with Time Windows (TOPTW) based on [6] benchmark set.

The paper is organized in the following way. Section 2 briefly overviews some related approaches. Section 3 states the notation and formally defines the problem. The proposed method is described in Sect. 4. Extensive computational experiments are reported in Sect. 5. Section 6 concludes the paper.

2 Related Work

The crucial point of our Deep Infeasibility Exploration Method (DIEM) is the appropriate handling of constraints in the VRP. Various constraints need to be taken into account also in many other NP-hard optimization problems. For example, feasible solutions to the knapsack problem need to meet the knapsack capacity constraint, or several capacity constraints in the case of the multidimensional knapsack problem [12]. The solutions to scheduling problems need to respect various precedence constraints [1]. At the same time, most metaheuristic algorithms are designed for unconstrained optimization [23]. Thus, some additional constraint handling techniques are necessary to solve hard constrained optimization problems, including VRPs, with metaheuristics.

General constraint handling techniques used in the context of metaheuristics may be classified as (see e.g. [23] Chapter 1.5, [5,14] for reviews):

– *Problem reformulation*, i.e. changing the representation of solutions and/or operators, e.g. neighborhood or recombination operators, such that the metaheuristic method may treat the problem as an unconstrained one. For example, the travelling salesperson problem is a constrained problem if the solutions are encoded as sets of edges; however, if the solutions are encoded as permutations of nodes, then each permutation defines a feasible solution and the problem may be treated as an unconstrained problem. This approach, however, may be used only for relatively simple constraints.

– *Rejection*, i.e. only feasible solutions are preserved, and each new infeasible solution is automatically rejected.
– *Repairing*, i.e. additional repair procedures are applied to each infeasible solution produced by the operators. A repair procedure is a heuristic that tries to restore feasibility with the minimum change of the solution. For example, in the case of the knapsack problem, a solution that exceeds the knapsack capacity may be repaired by removing superfluous items with the lowest value-to-weight ratios.
– *Decoding*, i.e. a metaheuristic encodes solutions using some indirect representation in which all solutions are treated as feasible and are decoded to the original representation when necessary. This approach may be used e.g. for scheduling problems, where the indirect representation may be a priority list of tasks and the decoding procedure may be a priority heuristic that constructs the final schedule based on the priority list. This approach is similar to problem reformulation, but differs from the latter by explicit use of two kinds of solutions representations.
– *Multi-objective approach*, in which both the original objective function and the level of constraints violation are treated as objectives and the problem is solved as a multiobjective one.
– *Penalty function*, i.e. some penalty terms dependent on the level of constraints violation are added to the objective function.

Approaches that work with feasible solutions only, like rejection or decoding, may seem highly favorable, but they may have a disadvantage of changing the objective function landscape. Formally, objective function landscape is defined as triplet (S, f, d), where S is a set of all solutions, $f : S \to \mathbb{R}$ is the objective function, and $d : S \times S \to \mathbb{R}$ is a distance or similarity measure that endows S with a structure (sometimes referred to as locality), and so turns it into a space. Since rejection may remove many solutions from the landscape, it may lose some desired properties, for example, the connectivity which makes each solution obtainable from another one with a series of simple transformations. Because of this risk, approaches that allow solution infeasibility, like penalty function, often prove to be more beneficial. Furthermore, the penalty function approach may be used with any kind of constraints, even with very complicated ones.

Penalty function approaches combine the original objective with some penalty term(s), usually involving carefully tuned weights [9]. The weights setting may be static, dynamic, i.e. changing in some predefined way, or adaptive. For example, fixed penalty weights for capacity and time window constraints for the VRPTW were applied in the penalty-based memetic algorithm of [16].

Although multiple versions of adaptive penalty functions have been proposed, the typical approach is to increase the penalty if the current solution(s) is infeasible, and decrease the penalty if the current solution(s) is feasible. Examples of this approach include a relaxation of capacity and time violation penalty weights in a tabu search metaheuristic for the multi-vehicle dial-a-ride problem [7], a vehicle overloading penalty adjustment for the VRP with backhauls [4], an adaptive adjustment of penalty weights for route duration, vehicle load and time window constraints in a genetic algorithm for a wide class of the VRPTW problems [26], or an adaptive mechanism to control the capacity and time window constraints in a path relinking approach for the VRPTW [11]. This usually results in a kind of oscillation of solutions in the proximity of the border between feasible and infeasible solutions. This behaviour conforms with the concept of strategic oscillation of [10], according to which good solutions are known to be close to the borders of feasibility. On the other hand, many infeasible solutions being further from the border of feasibility remain practically unavailable and the remaining part of the fitness landscape may still be difficult to explore by a metaheuristic.

In contrast to most of the works reviewed above, DIEM is designed to deeply explore the infeasible region. It allows even highly infeasible solutions to be produced when they are better in terms of the objective function without the penalty component. Whenever such solutions are found, we disturb them to create additional slackness, raise the penalty weights and try to make them feasible.

3   Problem Formulation

The general constrained optimization problem has the goal of finding a solution $s \in S$ which minimizes an objective function $f : S \to \mathbb{R}$ and is feasible.

We say that $s \in S$ is feasible if $p_i(s) = 0$ for each $i = 1, \ldots, N$, where each function $p_i : S \to \mathbb{R}_+ \cup \{0\}$ indicates the level of the i-th constraint violation and N is the number of constraints.

A vehicle routing problem consists in finding a set of possibly short routes for a homogeneous or heterogeneous fleet of vehicles, respecting some constraints dependent on the problem variant, e.g. with vehicle capacity limitation, and/or respecting time window constraints. Since even the simplest variants of VRPs are NP-hard [24], heuristic methods are usually used to solve them.

A specific VRP is defined by a set of depot nodes and a set of customer nodes. For each node, a non-negative service time, a pickup/delivery demand and a time window may be defined. For each pair of nodes (*edge*), a travel time and/or a travel distance are defined. A fleet of homogeneous or heterogeneous vehicles is available. The capacity of each vehicle might be limited. The goal is to find a set or routes that start and end at a depot, such that each customer node is visited exactly once, the sum of demands at each route does not exceed the vehicle capacity, and time windows are respected.

VRPs are usually formulated as bi-objective problems with two lexicographically ordered objectives. The first objective is to minimize the number of routes and the second one is to minimize the total travel distance. In experiments reported in this paper, we aggregate the two objectives with a weighted sum using a very large weight for the number of routes.

We also define an evaluation function which augments the objective function f with penalty terms:

$$\text{eval}(s, \vec{w}) = f(s) + \sum_{i=1}^{N} w_i p_i(s) \tag{1}$$

where $\vec{w} = [w_1, \ldots, w_N]$ is a vector of non-negative real-valued *penalty weights*. Depending on the VRP variant, we apply the following rules for determining the level of constraint violation:

- The capacity violation level is expressed as the sum of demands exceeding the vehicle capacities for individual routes, e.g. in VRPTW and PDPTW.
- The time window violation level is expressed as the sum of time units exceeded with respect to the time windows defined for particular customer nodes, e.g. in VRPTW, PDPTW and TOPTW.
- The node precedence violation level is expressed as the number of customer node pairs (pickup-delivery pairs) serviced by different vehicles or visited in wrong order (the delivery node is visited before the pickup node), e.g. in PDPTW.
- The route number violation level is expressed as the number of routes exceeding the limit of vehicles available, e.g. in TOPTW; for heterogeneous fleet, such numbers are calculated for each set of routes that have assigned a particular vehicle type, and next summed over all vehicle types.

4   Description of the Method

Algorithm 1 shows the complete pseudocode of DIEM. Each pool (main and working) owns a vector of penalty weights (`weights` attribute). These vectors are used to compare solutions and determine which is better (has lower eval value – see Eq. (1)).

```
1   working_pool ← generate random initial solutions
2   working_pool.weights ← improvement_weights
3   main_pool.weights ← 2 * improvement_weights
4   foreach s ∈ working_pool:
5       main_pool.append_or_replace(s)
6   prev_quality ← ∞
7   phase_type ← improvement
8   while not time exhausted:
9       do:
10          let underlying metaheuristics work on working_pool with
                working_pool.weights
11      while underlying metaheuristics make sufficient progress
12      foreach s ∈ working_pool:
13          main_pool.append_or_replace(s)
```

14   success ← (main_pool.quality() < prev_quality)
15   adjust main_pool.weights by Algorithm 2 (page 70)
16   prev_quality ← main_pool.quality()
17   adjust improvement_weights by Algorithm 3 (page 70)
18   **if** phase_type = *feasibility* **or** at least one solution in working_pool
            is feasible **or** $\min_{s \in \text{main\_pool}} f(s) \leq \min_{s \in \text{working\_pool}} f(s)$:
19   solutions ← main_pool
20   working_pool.weights ← improvement_weights
21   phase_type ← *improvement*
22   **else**: *//better found but infeasible*
23   solutions ← working_pool
24   working_pool.weights ← main_pool.weights
25   phase_type ← *feasibility*
26   working_pool.clear()
27   **foreach** s ∈ solutions:
28   working_pool.append_or_replace(ruin(s))
29 the best feasible solution in main_pool is the result

Algorithm 1. Deep Infeasibility Exploration Method (DIEM)

In the beginning (lines 2-3) we assign some penalty weights to both pools. The working pool is initialized with solutions generated by some randomized heuristics (line 1). The solutions are then inserted into the main pool (lines 4-5). The main loop of the algorithm (lines 8-28) interlaces the improvement and feasibility phases (according to the current value of **phase_type** variable). Each phase performs a search for better solutions using underlying metaheuristics working on the working pool in relatively short cycles (and using appropriate penalty weights to evaluate solutions). The metaheuristics like large neighborhood search, iterated local search, or evolutionary algorithms work concurrently, drawing solutions from the working pool, trying to improve them and to insert the improved versions again into the pool (more details on the insertion algorithm are given later in this section). Each metaheuristic can be run simultaneously in multiple threads, accordingly to computational resources available. We allow underlying methods to work as long as they make progress (lines 9-11). Strictly speaking, we require the methods to make a minimal progress (expressed as percentage of the current best solution cost, e.g. 0.1%) in a predefined time interval.

After each search phase, solutions stored in the working pool are inserted into the main pool (lines 12-13). If the last phase finds solutions which improve the quality of the main pool, the **success** flag is set in line 14. The quality of the pool is evaluated by **main_pool.quality()** calls in lines 14 and 16. This function calculates an arithmetic mean over eval(s, **main_pool.weights**), where s iterates over $\lfloor L/2 \rfloor$ best solutions in the pool, L is the pool capacity ($L = 32$ in our case), and the eval function is defined by (1). We use only 50% best solutions in order to avoid influence of potential outlier solutions much worse and different from the best solutions in the pool. In the next steps we adjust the penalty weights (lines 15 and 17), which is described in detail later in this section.

In line 18 we decide which phase should be run next and which solutions should be (after disturbing) used as starting solutions for this phase: those from the working pool or those from the main pool.

After deciding which solutions should be used in the subsequent phase, we replace (in lines 26-28) the content of the working pool with disturbed (by `ruin` function) versions of them. In general, more powerful underlying metaheuristics require stronger disturbance operations, or else they tend to return to previous solutions. It is natural to make disturbance level adaptive as it is domain specific. For the VRP family, replacing $10-30\%$ of the edges is sufficient to avoid returning to starting solutions, yet to preserve some structure of the solutions. We apply the Adaptive Large Neighborhood Search ruin operators of [19] with the level of destruction defined as the number of nodes to be removed from a solution and then reinserted according to the regret reinsertion heuristic. This level of solution destruction is adaptively changed in subsequent phases of the algorithm. It starts with the maximal value $\min(6\sqrt{P}, P/2)$ (chosen empirically), where P is the instance size defined as the total number of nodes. If at least one of the solutions to be ruined turns out identical as produced in the previous phases, the destruction level is increased; otherwise, it is decreased down to $\sqrt{P}$.

DIEM uses two pools of solutions. The content of each pool is internally stored in two sets:

1. R – a set of solutions with limited capacity (in the experiments presented in this paper we apply for both pools the same limit of 32 solutions);
2. F – an empty set or a singleton whose element is the best feasible solution added to the pool. F is empty only if no feasible solution has been added to the pool yet.

Additionally, each pool owns a vector of penalty weights (`weights` attribute). The procedures for adding and evaluation use this vector of weights to compare solutions and determine which is better (has lower eval value – see Eq. (1)).

To ensure diversity of solutions stored in each pool we use a reflexive and symmetric predicate sim : $S \times S \rightarrow \{$true, false$\}$ which determines whether two solutions are similar. In our case the predicate checks if the size of symmetric difference between sets of undirected edges of the arguments is lower than a threshold. The threshold depends on the instance size P being the total number of nodes, including depot and customer ones. We empirically set the threshold value to $\min(6\sqrt{P}, P/2)$.

The algorithm of adding a solution s to a pool (`append_or_replace` method) works as follows:

1. If s is feasible and F is empty or s is better than the solution in F, then s replaces the content of F.
2. If $s \notin R$, then an attempt is made to add it there.
   Let Z be a subset of R consisting of all solutions similar to s according to sim predicate:
   - If Z is empty, then s is added to R if it does not violate the capacity limit or if it is better than the worst solution included in R. The worst solution is removed from R if the capacity limit has been exceeded.

– If Z is not empty, then s is added to R only if it is better than the best solution in Z and then the whole Z is subtracted from R.

Note that the above addition strategy guarantees that no two solutions in R are similar.

The main pool stores best (almost) feasible solutions obtained so far by the algorithm. Therefore, the weights used by this pool have to be adjusted in a conservative way. Changing them too much may have detrimental effect on the performance – decreasing the weights too much is likely to flood the pool with highly infeasible solutions, while increasing them excessively may cause an influx of very weak solutions. The idea of weight modification in this pool is similar to that of [11], but with an important difference: we adjust the weights after multiple runs of underlying metaheuristics, while [11] adjusted their weights after every single run of local search. As our adjustment is less frequent, we use more aggressive multipliers (compared to 0.99 and 1.01 used in the cited work).

```
1   if phase_type = feasibility:
2       if the best solution in working_pool is feasible:
3           decrease main_pool.weights by factor D^main
4       else:
5           increase violated main_pool.weights by factor I^main
```
Algorithm 2. Adjusting penalty weights of the diversified pool.

Algorithm 2 shows in detail the process of adjusting weights of the main pool (`main_pool.weights`). The weights are changed only after feasibility restoration phase (see line 1) since they are used only there. The algorithm decreases (line 3) or increases (line 5) the weights depending on the feasibility of the best solution in the working pool (line 2). Strictly speaking, only the weights of constraints violated by the best solution in the working pool are increased in the latter case.

The improvement weights (i.e. the weights used by the working pool in improvement phases) are adjusted according to Algorithm 3. These weights are adjusted in each iteration of the main loop of DIEM. Since we do not expect the improvement phase to produce feasible solutions directly, we can use relatively low improvement weights. On the other hand, the feasibility restoration phase should be able to boost their feasibility enough to improve them in terms of (high) weights of the main pool. Therefore, the improvement weights cannot be too low and too different from weights of the main pool. That is why we increase them only after feasibility restoration phase (in line 5) and this is done only if solutions in the main pool have not been improved in terms of `main_pool.weights` (in that case, the `success` flag introduced in Algorithm 1 and tested in line 2 is not set). In the opposite case, the weights are decreased (line 3). The weights are also decreased after the improvement phase, in two cases: when the working pool includes at least one feasible solution (lines 7-8) or when the quality of the best solution in the working pool is not better than its counterpart in the main pool (lines 9-10).

```
1   if phase_type = feasibility:
2       if success:
3           decrease violated improvement_weights by factor D_1^{imp}
```

4   **else:**
5   increase violated improvement_weights by factor I^{imp}
6   **else:**
7   **if** working_pool includes at least one feasible solution:
8   decrease improvement_weights by factor D_2^{imp}
9   **else if** $\min_{s \in \text{main\_pool}} f(s) \leq \min_{s \in \text{working\_pool}} f(s)$:
10   decrease violated improvement_weights by factor D_3^{imp}

Algorithm 3. Adjusting improvement weights.

In the improvement phase, the goal is to dive deep into infeasibility as fast as possible. Therefore, if feasible solutions have been obtained after the improvement phase, we significantly decrease the improvement weights (in line 8). The results in the working pool are temporary and thus a less conservative approach than in the case of the main pool is appropriate. After reaching the infeasible regions, our aim is to find solutions which are better in terms of the objective function. If this is not achieved, we still decrease the weights of violated constraints (in line 10).

In the feasibility restoration phase, a success is defined as finding „repairable" solutions better in terms of the objective function. If the success is achieved we slightly decrease the weights of violated constraints (in line 3) - we aim to find as low „repairable" weights as possible to broaden the space searched by the underlying metaheuristics. If feasibility restoration phase was not successful, we increase the weights for the infeasible dimensions of the problem (in line 5).

5   Computational Experiments

The goal of the computational experiment is twofold. On the one hand, to compare DIEM to the state-of-the-art results presented in the literature. On the other hand, to show that the proposed penalty weights and diversity management method significantly improves the performance of DIEM in comparison to the standard adaptive penalty oscillation with the same set of underlying metaheuristics.

DIEM was tested on the standard set of benchmark instances by [8]. This set consists of VRPTW instances with a homogeneous fleet, divided into six groups. Groups c1 and c2 include the customers located in clusters, while in groups r1 and r2 the locations are generated completely randomly. Groups rc1 and rc2 contain a mix of completely random and clustered customers. Groups c1, r1 and rc1 have smaller vehicle capacities and shorter time windows than groups c2, r2 and rc2. The benchmark consists of five subsets containing 200, 400, 600, 800 and 1000 customer nodes, with 60 instances in each subset (10 instances in each group c1, c2, r1, r2, rc1, and rc2), resulting in 300 instances.

We compared DIEM to the well-known state-of-the-art methods published in the literature with competitive best-known results for the [8] instances: the method of Nagata et al. [16] and the method of Vidal et al. [26] using results published in these papers. Note that these methods use relatively simple penalty

Table 1. Summarized results for the VRPTW [8] benchmark instances. First six rows for each subset of the instances are in the format "Fleet size | Distance".

instances		Nagata et al.	Vidal et al.	DIEM avg 5	DIEM best 5	HY-oscillation
200	c1	18.9 | 2718.44	**18.9 | 2718.41**	18.9 | 2718.41	**18.9 | 2718.41**	18.9 | 2718.41
	c2	6.0 | 1831.64	**6.0 | 1831.59**	6.0 | 1831.59	**6.0 | 1831.59**	6.0 | 1831.59
	r1	18.2 | 3615.15	18.2 | 3613.16	18.2 | **3611.93**	18.2 — 3611.93	18.2 | 3611.93
	r2	4.0 | 2930.04	**4.0 | 2929.41**	4.0 | 2929.41	**4.0 | 2929.41**	4.0 | 2929.41
	rc1	18.0 | 3182.48	18.0 | 3180.48	18.0 | **3176.23**	18.0 | 3176.23	18.0 | 3176.23
	rc2	4.3 | 2536.54	4.3 | 2536.20	4.3 | **2535.88**	4.3 | 2535.88	4.3 | 2535.88
	CNV	694	694	694	694	694
	CTD	168 143	168 092	168 035	168 035	168 035
	Time	4.7	5 × 8.4	5 × 10	5 × 10	5 × 10
400	c1	37.6 | 7179.71	37.6 | 7170.47	37.6 | 7176.74	**37.6 | 7170.31**	37.7 | 7211.47
	c2	11.7 | 3898.02	11.6 | 3952.95	11.6 | 3951.40	**11.6 | 3943.18**	11.7 | 3911.57
	r1	36.4 | 8413.23	36.4 | 8402.57	36.4 | 8399.08	**36.4 | 8393.60**	36.4 | 8493.45
	r2	8.0 | 6149.49	8.0 | 6152.92	8.0 | 6152.29	**8.0 | 6149.39**	8.0 | 6186.95
	rc1	36.0 | 7931.66	36.0 | 7907.14	36.0 | 7897.26	**36.0 | 7893.90**	36.0 | 7965.63
	rc2	**8.4 | 5293.74**	8.5 | 5215.21	8.5 | 5213.30	8.5 | 5210.93	8.5 | 5282.35
	CNV	1381	1381	1381	1381	1383
	CTD	388 548	388 013	387 901	387 747	390 514
	Time	34.0	5 × 34.1	5 × 60	5 × 60	5 × 60
600	c1	57.4 | 14054.70	57.4 | 14058.46	57.4 | 14041.73	**57.4 | 14038.81**	57.4 | 14115.60
	c2	17.4 | 7601.94	17.4 | 7594.41	17.4 | 7593.27	**17.4 | 7582.92**	17.4 | 7667.21
	r1	54.5 | 18194.38	54.5 | 18023.18	54.5 | 18004.58	**54.5 | 17992.32**	54.5 | 18221.36
	r2	**11.0 | 12319.75**	11.0 | 12352.38	11.0 | 12339.70	11.0 | 12334.04	11.0 | 12525.35
	rc1	55.0 | 16179.39	55.0 | 16097.05	55.0 | 16072.28	**55.0 | 16057.09**	55.0 | 16216.42
	rc2	**11.4 | 10591.87**	11.5 | 10511.86	11.5 | 10523.50	11.5 | 10505.54	11.5 | 10721.53
	CNV	2067	2068	2068	2068	2068
	CTD	789 420	786 373	785 751	785 617	794 675
	Time	80.4	5 × 99.4	5 × 120	5 × 120	5 × 120
800	c1	75.3 | 24990.42	75.4 | 24876.38	75.3 | 24923.48	**75.3 | 24898.59**	75.3 | 25043.68
	c2	23.4 | 11438.52	23.3 | 11475.05	23.2 | 11585.78	**23.2 | 11577.48**	23.4 | 11550.11
	r1	72.8 | 31486.74	72.8 | 31311.38	72.8 | 31298.73	**72.8 | 31283.19**	72.8 | 31557.88
	r2	15.0 | 19873.04	15.0 | 19933.39	15.0 | 19889.48	**15.0 | 19871.48**	15.0 | 20152.31
	rc1	72.0 | 31020.22	72.0 | 29404.32	72.0 | 29301.43	**72.0 | 29279.22**	72.0 | 29523.94
	rc2	15.4 | 16438.90	15.4 | 16495.82	15.4 | 16459.40	**15.4 | 16437.49**	15.5 | 16600.05
	CNV	2739	2739	2737	2737	2739
	CTD	1 352 478	1 334 963	1 334 583	1 333 749	1 344 280
	Time	126.8	5 × 215	5 × 240	5 × 240	5 × 240
1000	c1	94.1 | 41683.29	94.1 | 41572.86	94.1 | 41540.38	**94.1 | 41517.72**	94.2 | 41596.20
	c2	29.1 | 16498.61	28.8 | 16796.45	28.8 | 16707.63	**28.8 | 16698.66**	29.1 | 16485.10
	r1	91.9 | 48287.98	91.9 | 47759.66	91.9 | 47644.13	**91.9 | 47603.00**	91.9 | 47986.37
	r2	**19.0 | 28913.40**	19.0 | 29076.45	19.0 | 28960.99	19.0 | 28936.23	19.0 | 29254.49
	rc1	90.0 | 44743.18	90.0 | 44333.40	90.0 | 44322.46	**90.0 | 44277.98**	90.0 | 44675.11
	rc2	18.3 | 23939.62	18.2 | 24131.13	18.2 | 24105.69	**18.2 | 24075.52**	18.2 | 24378.61
	CNV	3424	3420	3420	3420	3424
	CTD	2 040 661	2 036 700	2 032 813	2 032 154	2 043 759
	Time	186.4	5 × 349	5 × 360	5 × 360	5 × 360

management, i.e. constant penalty weights in the case of [16] and adaptive penalty oscillation in the case of [26].

Additionally, we included in the experiment our own implementation of the well-known penalty oscillation method introduced by [11] for the VRPTW (referred to as HY-oscillation) with the same set of underlying metaheuristics that used in the implementation of DIEM. HY-oscillation uses a single pool of solutions without disturbing the entire population before starting a new optimization phase. The method adapts penalty weights in a way similar to [26], i.e. increases the penalty weights of violated constraints (capacity or time window violations) if the current best solution is infeasible, and decreases them if the current best solution is feasible.

The underlying metaheuristics used in both DIEM and the HY-oscillation, i.e. large neighborhood search [19], iterated local search [25], and evolutionary algorithms [16,26], were implemented by ourselves following their descriptions in previously published papers.

Both our method and HY-oscillation were given the same initial starting weights and computing time similar to that used by [26]. Weight adjustment factors used in Algorithms 2 and 3 have been found empirically and set to $D^{main} = 0.96, I^{main} = 1.05, D_1^{imp} = 0.96, D_2^{imp} = 0.85, D_3^{imp} = 0.97, I^{imp} = 1.1$. Each instance was assigned a machine with two Intel Xeon E5-2697 v3 CPUs (14 simultaneous threads of execution each, resulting in 28 simultaneous threads of the underlying metaheuristics working concurrently).

The results of the comparison are shown in Table 1. The average and the best results of five runs of DIEM per instance are compared against those obtained by [16,26] and the HY-oscillation method. For brevity, we report only average results by group of instances (c1, c2, r1, r2, rc1, rc2) and different number of customer nodes (200, 400, 600, 800 and 1000) in the format "Fleet size | Distance". The best results are highlighted. The cumulative number of vehicles (CNV), the cumulative travel distance (CTD) and calculation time (Time) in minutes are also reported.

Instances with 200 customers proved to be trivial for our underlying metaheuristics, and both DIEM and HY-oscillation achieved verifiable best-known results according to the ranking on the [22] site. However, thanks to this experiment we were able to supplement this list with 31 solutions to instances with 200 customers which had previously lacked them (the best-known results for the instances had been published earlier by their authors without solutions provided). For the instances with 400, 600, 800 and 1000 customers DIEM was able to outperform the competitive methods in almost all groups of instances (20 out of 24). Our method achieved better results than the method of [16] except for 2 groups of r2 instances (600 and 1000 customers) and another 2 groups of rc2 instances (400 and 600 customers). Our results are better than the results obtained by [26] in all 24 groups. The cumulative travel distance of all instances is better, too.

The results indicate also that the same set of underlying metaheuristics when used without DIEM yields much worse results (see DIEM vs. HY-oscillation). Indeed, the results of HY-oscillation with the same underlying metaheuristics are often worse than the results of [16] and [26].

To check the continuous ability in finding good solutions, we performed multiple long-term (up to 24 hours) runs of DIEM for the [8] benchmark instances. In these experiments DIEM improved 117 out of 300 best-known solutions for the instances (marked on [22] site with SCR acronym, as of November 1, 2021). The highest improvements have been achieved mainly for the largest instances with 800-1000 customers, even by more than 2%, e.g. 2.19% for the 8th 800-customer instance in group c1. The detailed results are available upon request.

To prove the generality of our method, we performed tests on well-known benchmark datasets for two other classical vehicle routing problem variants: PDPTW benchmark instances by [13], and TOPTW datasets by [18] and [15].

The PDPTW has the same objective as the VRPTW but customer nodes are paired together into pickup-delivery tasks and a pair must be serviced by the same vehicle. Therefore, besides the capacity and time windows constraints, there are additionally precedence constraints to ensure that the pickup node is visited before the delivery node. The [13] benchmark dataset contains 300 PDPTW instances with 200, 400, 600, 800 and 1000 pickup and delivery tasks divided into six groups as in the [8] benchmark dataset (named similarly lc1, lc2, lr1, lr2, lrc1 and lrc2). We conducted multiple runs of DIEM giving each instance 1 to 24 ours of computing time accordingly to the instance size. During these experiments DIEM improved 99 out of 300 best-known solutions for the instances (marked on [21] site with SCR acronym, as of November 1, 2021). The highest improvements have been achieved again for the largest instances with 600-1000 tasks often by more than 1%, surprisingly even by 6.79% for the 7th 800-task instance in group lc2. It is worth mentioning that DIEM was able to improve also smaller instances (4 instances with 200 tasks) and some instances that failed to be improved for years, e.g. the 9th 800-task instance in group lr2 with the best-known result unbeaten since 2003. The detailed results are available upon request.

In TOPTW, each customer has assigned a score that is collected when customer is visited. The objective is to maximize the total collected score by visiting a set of customers with a limited number of routes and time window constraints. The benchmark instances for the TOPTW introduced by [18] (pr01-pr10) and [15] (pr11-pr20) have been adapted from Cordeau's instances [6]. The dataset consists of 20 instances with 48-288 customer nodes, optimized with different limitations of admissible route numbers $m \in \{2, 3, 4\}$, resulting in fact in 60 problem instances. We performed multiple runs of DIEM, giving each instance 1 to 24 hours of computing time accordingly to the instance size. In these experiments DIEM matched or improved 55 and strictly improved 39 out of 60 best-known solutions for the instances. The highest improvements have been achieved for the largest instances with 144-288 customers and for the highest number of admissible routes, 3 and 4, e.g. by almost 3.3% for the instance pr20 with 4 routes. Detailed results obtained by DIEM for particular instances and different route number limits are given in Table 2, where BK columns contain the current best-known results published on [17] site. All results generated by our method which match or improve the best-known solutions are highlighted (55 instances). The underlined ones have been achieved by DIEM as new best-known solutions (39 instances). The detailed results are available upon request.

Table 2. Detailed best results obtained with DIEM for the TOPTW [18] (pr01-pr10) and [15] (pr11-pr20) benchmark instances. Best-known solutions are highlighted (new best-known results provided by DIEM are underlined).

instance	size	$m = 2$		$m = 3$		$m = 4$	
		BK	DIEM	BK	DIEM	BK	DIEM
pr01	48	502	**502**	622	**622**	657	**657**
pr02	96	715	**715**	945	**945**	1079	**1081**
pr03	144	742	**742**	1011	**1013**	1233	**1245**
pr04	192	926	**928**	1294	**1298**	1585	**1595**
pr05	240	1101	**1101**	1482	**1498**	1838	**1854**
pr06	288	1076	**1076**	1514	**1516**	1860	**1892**
pr07	72	566	**566**	744	**744**	876	**876**
pr08	144	834	**834**	1139	**1142**	1382	**1394**
pr09	216	909	**909**	1275	**1282**	1619	**1623**
pr10	288	1134	**1145**	1573	**1586**	1943	**1970**
pr11	48	566	**566**	654	**654**	657	**657**
pr12	96	774	768	1002	**1004**	1132	**1136**
pr13	144	843	**846**	1152	**1155**	1386	**1393**
pr14	192	1017	995	1372	**1376**	1674	**1696**
pr15	240	1227	**1238**	1662	**1693**	2065	**2089**
pr16	288	1231	1199	1668	1660	2065	**2070**
pr17	72	652	646	841	**843**	934	**936**
pr18	144	953	**955**	1282	**1289**	1539	**1555**
pr19	216	1041	**1042**	1417	**1435**	1760	**1782**
pr20	288	1241	**1249**	1690	**1713**	2062	**2130**

These experiments confirmed that DIEM method does not get stuck in local optima even in very long runs and produces continuously high quality diverse solutions throughout an entire run, even for runs lasting several days and involving thousands cycles of improvement and feasibility restoration. This holds for virtually all VRP variants considered here. It is worth mentioning that the [8] and [13] benchmarks serve as an arena of continuous competition not only for scientists but also for companies working on VRP optimization. Verified best-known results are continuously published on [22] and [21] sites.

6 Conclusions

We have presented DIEM, a new metaheuristic method for the family of vehicle routing problems based on the penalty function approach with a deep exploration of the infeasible region, and a new diversity management scheme using two pools of solutions with different penalty weights and large disturbances of all solutions.

DIEM was tested on several variants of the vehicle routing problem, including VRPTW, PDPTW, and TOPTW. Our method was able to generate multiple new best-to-date solutions for standard benchmark instances. The computational experiments presented in this paper show that DIEM performs much better than the standard adaptive manipulation of the penalty weights as proposed by [11], with the same set of underlying metaheuristics used. The adjustment factors used in our method have been found empirically and systematic testing of sensitivity to their selection is one of the directions for further research.

Although we have tested the proposed method only on variants of the vehicle routing problem, the concept of deep infeasibility exploration and its implementation as the improvement and feasibility restoration phases is fairly general and could be adapted to other hard optimization problems with constraints. Of course, further computational experiments are needed to confirm the usefulness of DIEM in solving other problems. Thus, application of the proposed approach to other problems is a natural direction for further research. Another interesting direction for further research is comparison with other constraint-handling techniques that have not been used for VRPs, e.g. [20].

Acknowledgments. The authors would like to thank Krzysztof Krawiec for valuable comments on the paper.

The computing time for the reported experiments was provided to us by the generosity of Poznan Supercomputing and Networking Center, which had awarded us a computational grant to access its Eagle cluster. (Poznan Supercomputing and Networking Center, computing grant 358 (https://wiki.man.poznan.pl/hpc/index.php?title=Eagle)).

References

1. Blazewicz, J., Kobler, D.: Review of properties of different precedence graphs for scheduling problems. Eur. J. Oper. Res. **142**(3), 435–443 (2002). https://doi.org/10.1016/S0377-2217(01)00379-4
2. Blocho, M., Nalepa, J.: LCS-based selective route exchange crossover for the pickup and delivery problem with time windows. In: Hu, B., López-Ibáñez, M. (eds.) Evolutionary Computation in Combinatorial Optimization, pp. 124–140. Springer International Publishing, Cham (2017)
3. Braekers, K., Ramaekers, K., Nieuwenhuyse, I.V.: The vehicle routing problem: State of the art classification and review. Comput. Ind. Eng. **99**, 300–313 (2016). https://doi.org/10.1016/j.cie.2015.12.007
4. Brandao, J.: A new tabu search algorithm for the vehicle routing problem with backhauls. Eur. J. Oper. Res. **173**, 540–555 (2006). https://doi.org/10.1016/j.ejor.2005.01.042
5. Coello, C.A.C.: Theoretical and numerical constraint-handling techniques used with evolutionary algorithms: a survey of the state of the art. Comput. Meth. Appl. Mech. Eng. **191**(11), 1245–1287 (2002). https://doi.org/10.1016/S0045-7825(01)00323-1

6. Cordeau, J.F., Gendreau, M., Laporte, G.: A tabu search heuristic for periodic and multi-depot vehicle routing problems. Networks **30**(2), 105–119 (1997). https://doi.org/10.1002/(SICI)1097-0037(199709)30:2⟨105::AID-NET5⟩3.0.CO;2-G

7. Cordeau, J.F., Laporte, G.: A tabu search heuristic for the static multi-vehicle dial-a-ride problem. Transp. Res. Part B: Methodol. **37**(6), 579–594 (2003)

8. Gehring, H., Homberger, J.: A parallel hybrid evolutionary metaheuristic for the vehicle routing problem with time windows. In: University of Jyvaskyla, pp. 57–64 (1999)

9. Glover, F.: Future paths for integer programming and links to artificial intelligence. Computer. Oper. Res. **13**(5), 533–549 (1986). https://doi.org/10.1016/0305-0548(86)90048-1

10. Glover, F., Hao, J.K.: The case for strategic oscillation. Ann. Oper. Res. **183**(1), 163–173 (2011). https://doi.org/10.1007/s10479-009-0597-1

11. Hashimoto, H., Yagiura, M.: A path relinking approach with an adaptive mechanism to control parameters for the vehicle routing problem with time windows. In: van Hemert, J., Cotta, C. (eds.) Evol. Comput. Combin. Optim., pp. 254–265. Springer, Berlin Heidelberg, Berlin, Heidelberg (2008)

12. Kellerer, H., Pferschy, U., Pisinger, D.: Knapsack Problems. Springer, Berlin (2004)

13. Li, H., Lim, A.: A metaheuristic for the pickup and delivery problem with time windows. In: Proceedings of the 13th IEEE International Conference on Tools with Artificial Intelligence. pp. 160. ICTAI 2001, IEEE Computer Society, Washington, DC, USA (2001)

14. Mezura-Montes, E., Coello, C.A.C.: Constraint-handling in nature-inspired numerical optimization: past, present and future. Swarm Evol. Comput. **1**(4), 173–194 (2011). https://doi.org/10.1016/j.swevo.2011.10.001

15. Montemanni, R., Gambardella, L.: Ant colony system for team orienteering problems with time windows. Found. Comput. Decis. Sci. **34**, 287–306 (2009)

16. Nagata, Y., Braysy, O., Dullaert, W.: A penalty-based edge assembly memetic algorithm for the vehicle routing problem with time windows. Comput. Oper. Res. **37**(4), 724–737 (2010). https://doi.org/10.1016/j.cor.2009.06.022

17. OPLIB: The Orienteering Problem Library (2018). https://unicen.smu.edu.sg/oplib-orienteering-problem-library/

18. Righini, G., Salani, M.: New dynamic programming algorithms for the resource constrained elementary shortest path problem. Networks **51**, 155–170 (2008). https://doi.org/10.1002/net.20212

19. Ropke, S., Pisinger, D.: An adaptive large neighborhood search heuristic for the pickup and delivery problem with time windows. Transp. Sci. **40**, 455–472 (2006). https://doi.org/10.1287/trsc.1050.0135

20. Runarsson, T.P., Yao, X.: Stochastic ranking for constrained evolutionary optimization. IEEE Trans. Evol. Comput. **4**(3), 284–294 (2000). https://doi.org/10.1109/4235.873238

21. SINTEF's TOP PDPTW: TOP PDPTW Li & Lim benchmark (2019). https://www.sintef.no/projectweb/top/pdptw/li-lim-benchmark/

22. SINTEF's TOP VRPTW: TOP VRPTW Gehring & Homberger benchmark (2019). https://www.sintef.no/projectweb/top/vrptw/homberger-benchmark/

23. Talbi, E.G.: Metaheuristics: From Design to Implementation. Wiley Publishing, Hoboken (2009)

24. Toth, Paolo, Vigo, Daniele: Exact solution of the vehicle routing problem. In: Crainic, Teodor Gabriel, Laporte, Gilbert (eds.) Fleet Management and Logistics. CRT, pp. 1–31. Springer, Boston, MA (1998). https://doi.org/10.1007/978-1-4615-5755-5_1

25. Vaz Penna, P.H., Subramanian, A., Ochi, L.S.: An iterated local search heuristic for the heterogeneous fleet vehicle routing problem. J. Heuristics **19**(2, SI), 201–232 (2013). https://doi.org/10.1007/s10732-011-9186-y
26. Vidal, T., Crainic, T.G., Gendreau, M., Prins, C.: A hybrid genetic algorithm with adaptive diversity management for a large class of vehicle routing problems with time-windows. Comput. Oper. Res. **40**(1), 475–489 (2013). https://doi.org/10.1016/j.cor.2012.07.018

Evolutionary Algorithms for the Constrained Two-Level Role Mining Problem

Simon Anderer[1]([✉]), Falk Schrader[1], Bernd Scheuermann[1],
and Sanaz Mostaghim[2]

[1] University of Applied Sciences Karlsruhe, Karlsruhe, Germany
{Simon.Anderer,Bernd.Scheuermann}@h-ka.de
[2] Otto-von-Guericke Universität Magdeburg, Magdeburg, Germany
Sanaz.Mostaghim@ovgu.de

Abstract. The administration of access control structures in Enterprise Resource Planning Systems (ERP) is mainly organized by Role Based Access Control. The associated optimization problem is called the Role Mining Problem (RMP), which is known to be NP-complete. The goal is to search for role concepts minimizing the number of roles. Algorithms for this task are presented in literature, but often they cannot be used for role mining in ERP in a straightforward way, as ERP systems have additional conditions and constraints. Some ERP systems require multiple levels of roles. This paper defines new two-level variants of the RMP, examines their relationship and presents three approaches to computing such hierarchical role concepts. One is aiming at optimizing multiple levels of roles simultaneously. The other approaches divide the multi-level role mining problem into separate sub-problems, which are optimized individually. All approaches are based on an evolutionary algorithm for single-level role mining and have been implemented and evaluated in a range of experiments.

Keywords: Multi-level Role Mining · Evolutionary Algorithm ·
Access Control · Enterprise Resource Planning Systems

1 Introduction

Companies and other organizations use IT systems to efficiently support their business activities. Enterprise Resource Planning (ERP) systems, for example, support business processes at enterprise level comprising a wealth of different functions. The amount of sensitive data in such systems is growing considerably. Access control methods are used to ensure safe operations and compliance with rules and regulations and must be well-designed to ensure that users get

This work is supported by the German Ministry of Education and Research under grant number 16KIS1000.

L. Pérez Cáceres and S. Verel (Eds.): EvoCOP 2022, LNCS 13222, pp. 79–94, 2022.
https://doi.org/10.1007/978-3-031-04148-8_6

all permissions required to fulfill their job functions in the company. On the other hand, access control must prevent users from accidentally or deliberately causing damage. Access control in typical ERP systems is based on *Role Based Access Control (RBAC)* [17], which involves an indirect assignment of permissions to users: Each user receives a set of roles, where a role is defined as job function within an organization. Each role has a set of assigned permissions, where a permission describes the right to perform a particular operation on a certain protected business object (e.g. database record or file). To facilitate the administration of ERP systems, the goal is to minimize the number of roles, thereby considering the required user to permission assignment that is already known [12]. The corresponding optimization problem is called the *Role Mining Problem (RMP)* and was shown to be NP-complete [20]. SAP is the world leading provider of enterprise software. The software SAP ERP uses a more complex RBAC model, where roles are assigned in a two-level hierarchy. To meet this challenge, this paper defines two new variants of the RMP, the *Basic Two-level Role Mining Problem* and the *Constrained Two-level Role Mining Problem*, examines their relationship and presents three solution methods for the new problems.

2   Problem Description

A first definition of the RMP was given in [20] as minimum biclique cover problem. However, this paper introduces the RMP as binary matrix decomposition problem, using the following definitions:

- $U = \{u_1, u_2, ..., u_m\}$ a set of $m = |U|$ users,
- $P = \{p_1, p_2, ..., p_n\}$ a set of $n = |P|$ permissions,
- $R = \{r_1, r_2, ..., r_k\}$ a set of $k = |R|$ roles,
- $UPA \in \{0, 1\}^{m \times n}$ the targeted user to permission assignment matrix, where $UPA_{ij} = 1$ implies that permission p_j is assigned to user u_i,
- $UA \in \{0, 1\}^{m \times k}$ the matrix representing a possible assignment of roles to users, where $UA_{ij} = 1$ implies that role r_j is assigned to user u_i,
- $PA \in \{0, 1\}^{k \times n}$ the matrix representing a possible assignment of permissions to roles, where $PA_{ij} = 1$ implies that permission p_j is assigned to role r_i.

The *Basic Role Mining Problem* can now be defined based on these definitions. Given a set of users U, a set of permissions P and a user to permission assignment UPA, find a minimal set of roles R, a corresponding user to role assignment UA and a role to permission assignment PA, such that each user has exactly the set of permissions granted by UPA:

$$\textbf{Basic RMP} = \begin{cases} \min & |R| \\ \text{s.t.,} & \|UPA - UA \otimes PA\|_1 = 0. \end{cases} \tag{1}$$

where $\|.\|_1$ denotes the L_1-norm for matrices and $\otimes$ the Boolean Matrix Multiplication: $(UA \otimes PA)_{ij} = \bigvee_{l=1}^{k}(UA_{il} \wedge PA_{lj})$.

Figure 1 shows the different elements of a role concept $\pi = \langle R^\pi, UA^\pi, PA^\pi \rangle$, which denotes a candidate solution for a given Basic RMP, consisting of a set of roles R^π, a user to role assignment UA^π and a role to permission assignment PA^π. At this, black cells indicate 1's, white cells represent 0's. A candidate solution is called a feasible solution, if it satisfies the constraint in (1). For the Basic RMP, in particular, a feasible solution is also denoted *0-consistent*.

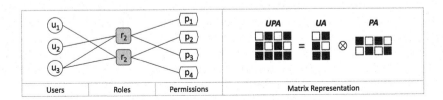

Fig. 1. Graph and matrix representation of the Basic RMP.

SAP ERP provides two levels of roles: so-called *single roles*, which represent rather small job functions in an organization, and *composite roles*, which correspond to large job functions or business processes, see Fig. 2. In SAP ERP, a set of permissions is assigned to each single role and a set of single roles is assigned to each composite role. Thus, permissions are not directly assigned to composite roles, but are inherited from the assigned single roles. A set of composite roles is assigned to each user. This can be considered a two-level role hierarchy and results in additional requirements for classical role mining.

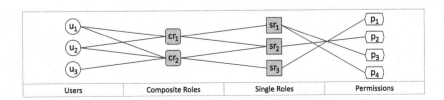

Fig. 2. Graph representation of role levels in SAP ERP.

3 Related Work

The Role Mining Problem is a well-studied problem in literature. Hence, there are many established solution techniques. A detailed survey on different RMP-variants and solutions strategies is provided by Mitra et al. [12], whereas the application of evolutionary algorithms for the RMP is considered in detail in [2]. Therefore, only the different solution strategies and the most important contributions are presented in the following. One common approach to tackle the RMP consists in permission grouping. At this, a set of roles is created from grouping permissions to roles. These roles are then assigned to users [4,10,14,18,21].

Another popular approach is based on mapping the RMP to other, well-known problems in data mining, applying compatible solution strategies to these problems [7,9,11,20]. Further approaches are based on formal concept analysis [13] or graph optimization [22]. Moreover, evolutionary algorithms were applied to the RMP [5,6,15]. In particular, in dynamically changing business environments evolutionary algorithms were used to mine for roles [1,16].

Schlegelmilch and Steffens (2005) were the first to apply clustering techniques to derive multi-level role concepts [18]. However, no overlap was allowed between roles in terms of shared permissions, leading to the creation of many hierarchy levels. Zhang et al. provide a graph-based approach in 2007 [22]. Roles are compared in pairs and arranged in a hierarchy. Permissions can be directly assigned to roles at all hierarchy levels. In 2010, this approach is extended in such a way that the deletion of obsolete roles becomes possible [23]. Guo et al. present a similar approach in 2008. At this, existing roles are iteratively included into a role hierarchy [8]. Molloy et al. (2008) as well as Takabi et al. (2010) use formal concept analysis [13,19]. In general, however, hierarchical role mining is more about finding an optimum role hierarchy with an unrestricted number of levels rather than an optimum two-level arrangement of composite and single roles.

4   Two-Level Role Mining

In this section the *Basic Two-level Role Mining Problem* and the *Constrained Two-level Role Mining Problem* are defined. For this purpose, the following additional notations are introduced:

- $S = \{sr_1, sr_2, ..., sr_{h_s}\}$ a set of $h_s = |S|$ single roles,
- $C = \{cr_1, cr_2, ..., cr_{h_c}\}$ a set of $h_c = |C|$ composite roles,
- $UCA \in \{0,1\}^{m \times h_c}$ a user to composite role assignment matrix,
- $CSA \in \{0,1\}^{h_c \times h_s}$ a composite role to single role assignment matrix,
- $SPA \in \{0,1\}^{h_s \times n}$ a single role to permission assignment matrix.

Based on that, the *Basic Two-level Role Mining Problem* (B2L-RMP) can be defined. Given a set of users U, a set of permissions P and a user to permission assignment UPA, find a set of composite roles C, a set of single roles S, a corresponding user to composite role assignment UCA, a composite role to single role assignment CSA and a single role to permission assignment SPA, that minimizes the total number of composite and single roles $|C| + |S|$, thereby ensuring that each user has exactly the set of permissions granted by UPA:

$$\text{B2L-RMP} = \begin{cases} \min & |C| + |S| \\ \text{s.t.,} & \|UPA - UCA \otimes CSA \otimes SPA\|_1 = 0. \end{cases} \tag{2}$$

The user to single role assignment matrix $USA = UCA \otimes CSA \in \{0,1\}^{m \times h_s}$ and the composite role to permission assignment matrix $CPA = CSA \otimes SPA \in \{0,1\}^{h_c \times n}$ can be obtained from Boolean matrix multiplication as illustrated in

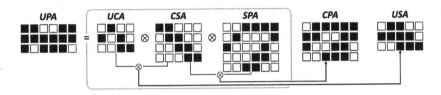

Fig. 3. Schematic representation of assignment matrices of the two-level RMP.

Fig. 3, which shows an example of the schematic representation of the different matrices and their interdependencies.

A two-level role concept $\varphi = \langle C^\varphi, S^\varphi, UCA^\varphi, CSA^\varphi, SPA^\varphi \rangle$ can be considered a possible solution to the B2L-RMP. The set of all possible solutions of the B2L-RMP is denoted by Φ. Again, if the constraint in (2) holds for a solution candidate, it is said to be *0-consistent*. It can be shown that the optimal solution of the B2L-RMP can be derived from the optimal solution of the Basic RMP:

Theorem 1. *Let r^* be the optimum number of roles of the Basic Role Mining Problem based on a set of users U, a set of permissions P and a user-permission assignment UPA. Then for the Basic Two-level RMP, based on the same U, P and UPA, the following holds:* $\min_{\varphi \in \Phi} (|C^\varphi| + |S^\varphi|) = 2r^*$.

Proof. Show that $\min_{\varphi \in \Phi} (|C^\varphi| + |S^\varphi|) \geq 2r^*$:

(a) For this, suppose there is a solution φ_1 of the Basic Two-level RMP with $|C^{\varphi_1}| < r^*$. Hence, $\pi_1 := \langle C^{\varphi_1}, UCA^{\varphi_1}, (CSA^{\varphi_1} \otimes SPA^{\varphi_1}) \rangle$ is a valid solution for the Basic RMP with $|R^{\pi_1}| = |C^{\varphi_1}| < r^*$ roles, which contradicts r^* being the minimum number of roles for the Basic RMP.

(b) Now, suppose there is a solution φ_2 of the Basic Two-level RMP with $|S^{\varphi_2}| < r^*$. Hence, $\pi_2 := \langle S^{\varphi_2}, (UCA^{\varphi_2} \otimes CSA^{\varphi_2}), SPA^{\varphi_2} \rangle$ is a valid solution for the Basic RMP with $|R^{\pi_2}| = |S^{\varphi_2}| < r^*$ roles. This again contradicts r^* being the minimum number of roles for the Basic RMP.

From (a) and (b), it follows directly that $|C^\varphi| + |S^\varphi| \geq 2r^*$ for all possible solutions $\varphi \in \Phi$ of the Basic Two-level Role Mining Problem.

Now, show that $\exists \varphi^* \in \Phi$ such that $|C^{\varphi^*}| + |S^{\varphi^*}| = 2r^*$:
Suppose that $\pi^* := \langle R^{\pi^*}, UA^{\pi^*}, PA^{\pi^*} \rangle$ is an optimum solution of the Basic RMP such that $|R^{\pi^*}| = r^*$. Then, $\varphi^* := \langle C^{\varphi^*}, S^{\varphi^*}, UA^{\pi^*}, I_{r^*}, PA^{\pi^*} \rangle$ is a possible solution of the Basic Two-level RMP, where the set of single roles S^{φ^*} corresponds to the set of roles R^{π^*}, hence $|S^{\varphi^*}| = |R^{\pi^*}| = r^*$. Furthermore, the i-th single role is assigned to exactly the i-th composite role only. Thus the composite role to single role assignment matrix corresponds to the r^*-dimensional identity matrix $CSA^{\varphi^*} = I_{r^*}$ and $|C^{\varphi^*}| = |R^{\pi^*}| = r^*$. Consequently, the total number of composite and single roles is $|C^{\varphi^*}| + |S^{\varphi^*}| = 2r^*$. $\qquad \square$

Theorem 1 does not only make statements about the number of composite and single roles of an optimum solution of the Basic Two-level RMP, but also shows how these roles can be constructed from an optimum solution π^* of the Basic RMP using trivial composite roles containing only one single role. It is obvious that this solution does not meet the requirements of ERP systems. For this reason, the Basic Two-level RMP is extended. In order to reflect the fact that single roles correspond to rather small job functions, an upper bound $s_{max} \in \mathbb{N}$ on the number of permissions contained in a single role is introduced. Based on the previous notations, the *Constrained Two-level RMP* (C2L-RMP), which will be considered throughout the remainder of this paper, can be defined as follows:

$$\text{C2L-RMP} = \begin{cases} \min & |C| + |S| \\ \text{s.t.,} & \|UPA - UCA \otimes CSA \otimes SPA\|_1 = 0, \\ & \sum_{j=1}^{n} SPA_{ij} < s_{max}, i \in \{1, ..., h_s\}. \end{cases} \quad (3)$$

5   Single-Level Optimization of Role Concepts

This section briefly introduces to the *addRole-EA*, an evolutionary algorithm to search for solutions to the Basic RMP. A detailed description can be found in [2]. Figure 4 provides a top-level description of the addRole-EA and lists the values of the parameters as adopted from [2]. One main characteristic of the addRole-EA is that at any time the 0-consistency constraint is satisfied.

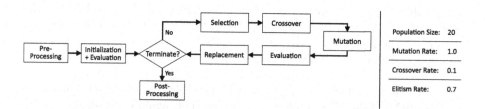

Fig. 4. Top-level description of addRole-EA incl. parameter values.

Pre-processing. To simplify the problem, four pre-processing steps have been identified, which are applied to the initial UPA matrix to reduce its dimension without loss of information. For example, users without any permission, or permissions, that are assigned to no user, are deleted.

Chromosome Encoding and Initialization. Analogous to the representation of the possible solutions of the Basic RMP $\pi = \langle R^\pi, UA^\pi, PA^\pi \rangle$, each individual comprises a set of roles R as well as a UA and a PA matrix. Based on a seed individual, where $UA = UPA$ and $PA = I_n$, an initial population is created by applying a random sequence of mutation operators to the seed individual. In order to save memory space, the sparse representation is used for the matrices of the individuals, see Fig. 5.

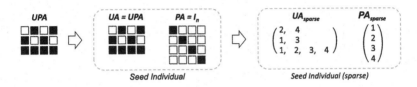

Fig. 5. Encoding of seed individual.

Evaluation of Fitness. As the addRole-EA is designed for the Basic RMP, the number of roles of an individual constitutes its fitness value.

The addRole-Method. The addRole-method is the principal method of the addRole-EA adding a new role to the set R. Accordingly, a row is added to the PA matrix. The new role is assigned to all users, having the corresponding permissions, by appending a new column to the UA matrix. In a second step, all roles that became obsolete by adding the given role are removed from R and the UA and PA matrices are updated. The 0-consistency constraint is therefore preserved and the resulting individuals are feasible solution candidates. One advantage of the addRole-Method consists in the fact that each role can be examined before its actual addition to an individual. Thus, it is possible to include further preferences, like the limitation of the number of permissions assigned to a role, at this point. Roles that do not comply with desired characteristics can already be rejected before the addRole-method is carried out.

Crossover and Mutation. The crossover method of the addRole-EA creates two child individuals from exchange of roles between two parent individuals. The different crossover methods differ in how the parent individuals and the roles for exchange are selected. The selected roles are added to the child chromosomes using the addRole-method, which ensures that only feasible solutions are created. In each mutation method, a new role is created and added to the individual by the addRole-method. The new roles are for example obtained from shared permissions of different users or from the union of the permissions assigned to different existing roles.

Replacement. The addRole-EA is a steady-state evolutionary algorithm. For replacement an elitist selection scheme is applied, which means that the best individuals among parents and offspring are selected.

Post-processing. At the end, the obtained role concepts must be extended to match the original problem specifications before pre-processing. For this purpose, the deletion of users and permissions, has to be undone.

6 Two-Level Optimization of Role Concepts

In this section, three approaches for the Constrained Two-level RMP are presented. First, the addRole-EA is executed twice to optimize composite and single roles consecutively. The second approach, is also based on the addRole-EA, where both role levels are optimized alternatingly. In the third approach, both role levels are optimized simultaneously. For this purpose, the addRole-EA is extended. For all approaches, the overall objective is the same: search for an optimum solution $\varphi^* = \langle C^{\varphi^*}, S^{\varphi^*}, UCA^{\varphi^*}, CSA^{\varphi^*}, SPA^{\varphi^*} \rangle$ of the C2L-RMP.

6.1   Consecutive Optimization of Role Levels

The straightforward way to obtain two-level role concepts is to divide the two-level role mining problem into two one-level sub-problems. These sub-problems are then solved in separate runs of the addRole-EA:

Variant 1: Single Roles First (SRF). The addRole-EA is used to compute single roles from the UPA matrix in a first run. The best individual of this run provides a user-single role assignment USA^* and a single role-permission assignment SPA^* as output. Subsequently, in a second run, the addRole-EA is used to compute composite roles from the USA^* matrix. The best individual of the second run then provides the composite-single role assignment CSA^* and the assignments of composite roles to users UCA^*, see Fig. 6 (left).

Variant 2: Composite Roles First (CRF). Composite roles are computed from the UPA matrix in a first run of the addRole-EA. Here, the best individual of the first run, provides UCA^* and CPA^* as output. Subsequently, single roles are mined from CPA^* in a second run, see Fig. 6 (right).

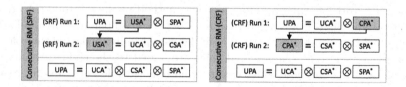

Fig. 6. Consecutive two-level role mining.

6.2   Alternating Optimization of Role Levels

The concept of alternating two-level role mining is based on the consecutive mining approach of the previous subsection. The two-level role mining problem is again divided into the same one-level sub-problems. The main difference compared to consecutive two-level role mining is, that it alternates between the optimization of single and composite roles, as it might be reasonable to re-optimize composite roles after the optimization of single roles and vice versa. For this purpose, the addRole-EA is applied alternatingly to the two sub-problems for a certain number of iterations. Analogous to (V2) of the consecutive case, composite roles are optimized using the addRole-EA in a first run. After optimization, C^*, UCA^* and CPA^*, can be obtained from the chromosome of the best individual. Subsequently, the addRole-EA is applied to the CPA^* matrix to optimize single roles in a second run, resulting in S^*, CSA^* and SPA^*. In the third run, again composite roles are optimized using $USA^* = UCA^* \otimes CSA^*$ as input matrix of the addRole-EA. The single roles in S^* and the single role to permission assignment SPA^* remain unchanged. This procedure is now continued iteratively, see Fig. 7. In contrast to the original version of the addRole-EA, the two matrices of the best individual of the previous run on the same role level are used for initialization of the seed individual instead of using the identity.

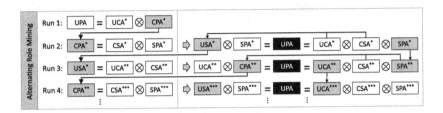

Fig. 7. Alternating two-level role mining.

6.3   Simultaneous Optimization of Role Levels

In consecutive and alternating two-level role mining, only one level of the overall problem is optimized at a time, while the other level is deliberately excluded from the optimization process during this time. In simultaneous role mining, role concepts, consisting of composite and single roles, are computed at once. The algorithm for simultaneous role mining, which is introduced in the following, is based on the addRole-EA. The general scheme (see Fig. 4) and typical parameters, like population size or crossover and mutation rate, are adopted from the single-level approach, but some steps are adapted:

Chromosome Encoding and Initialization. Analogous to the representation of the possible solutions of the C2L-RMP $\varphi = \langle C^\varphi, S^\varphi, UCA^\varphi, CSA^\varphi, SPA^\varphi \rangle$, a seed individual, where $UCA = UPA$, $CSA = I_n$ and $SPA = I_n$, is used to create an initial population by applying a random sequence of mutation operators. All matrices are again encoded using the sparse representation.

Fitness. Based on the definition of the C2L-RMP, the fitness of an individual is obtained as the sum of the number of its composite and its single roles $|C| + |S|$.

Classification. Before the application of one of the addRole-Methods of the simultaneous approach, a given set of permissions $\hat{P} \subseteq P$ needs to be classified. In case $|\hat{P}| \leq s_{max}$, a single role is created, which is assigned all permissions of $\hat{P}$ and added to the chromosome of an individual by the addSingleRole-method. In case $|\hat{P}| > s_{max}$, the simultaneous approach aims at creating a composite role, that inherits all permissions of $\hat{P}$, applying the addCompositeRole-method.

The addSingleRole-Method. A single role sr_{new}, based on a set of permissions $\hat{P}_s$, is added to the set S of an individual in two steps: First, a new row is added to SPA and a column is appended to CSA. The single role is assigned to all composite roles in C, whose set of inherited permissions form a super set of $\hat{P}_s$. If no composite role is found that fulfills this condition, a new composite role cr_{new} is created, to which sr_{new} is assigned as single role. This composite role is then assigned to all suitable users such that the 0-consistency constraint of the C2L-RMP remains fulfilled. This requires an update of C, UCA and CSA. In a second step, for each single role in $S \setminus \{sr_{new}\}$ and each composite role in $C \setminus \{cr_{new}\}$, it is analyzed, whether it is still needed to fulfill the 0-consistency constraint. If this is not the case, it is deleted from the individual. In the example in Fig. 8 (left), since the set of inherited permissions of none of the existing composite roles forms a super set of the permissions assigned to sr_{new}, a new composite role cr_5 is created in CSA and SPA and assigned to u_2 and u_3 in UCA (step 1). It can be observed that sr_5 and cr_4 are no longer necessary to provide each user with all required permissions and can thus be deleted from UCA, CSA and SPA (step 2).

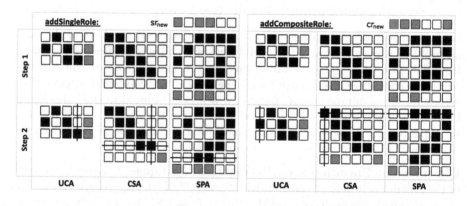

Fig. 8. Example of addSingleRole- and addCompositeRole-method.

The addCompositeRole-Method. Based on a given set of permissions $\hat{P}_c$, this method adds a new composite roles cr_{new} to the set C of an individual. Analogous to the design of the addSingleRole-method, two steps are required. First, a new row is added to CSA and a column is appended to UCA. Subsequently, cr_{new} is assigned to all users whose set of assigned permissions is a super set of $\hat{P}_c$. All single roles whose set of assigned permissions forms a subset of $\hat{P}_c$ are assigned to the cr_{new}. In case that not all permissions of the considered set of permissions $\hat{P}_c$ can be covered by existing single roles, a new single role sr_{new} is created and assigned the leftover permissions. This requires an update of S, CSA and SPA. In the second step, it is analyzed whether some of the composite roles of $C \setminus \{cr_{new}\}$ respectively some of the single roles in $S \setminus \{sr_{new}\}$ can be deleted. In Fig. 8 (right), cr_{new} is assigned to user u_2 in a first step. Furthermore, sr_2 is assigned to the new composite role. Since thereafter, p_1 and p_3 still remain uncovered, a new single role containing only these two permissions is created and assigned to cr_{new}. In the second step, it is found, that cr_1 and sr_1 are no longer needed. Thus, they are removed from UCA, CSA and SPA.

Crossover and Mutation. The crossover methods of the two-level algorithm operate analogously to the crossover methods of the addRole-EA. The only difference consists in the fact, that in each application of the crossover methods is decided randomly whether composite or single roles are selected for exchange. Also the mutation methods of the two-level algorithm can be obtained from the mutation methods of the addRole-EA in a straightforward manner. However, it must be noted that the resulting set of permissions must first be classified as single or composite role before applying one of the addRole-methods.

7   Evaluation

In this section, the presented methods for two-level role mining are evaluated. Since the C2L-RMP is a novel optimization problem, there are no other suitable solution strategies yet. Therefore, the evaluation aims less at conducting a detailed performance comparison but rather at showing the functionality of the developed methods. Evaluation was conducted on benchmark instances of the *PLAIN-benchmark* of RMPlib (*PLAIN_small_02* and *PLAIN_small_05*) [3]. It is reported that these instances in RMPlib were created by first generating random UA and PA matrices. The UPA matrix, which constitutes the benchmark instance, was then obtained from Boolean matrix multiplication $UPA = UA \otimes PA$ [3]. The instances of the RMPlib are therefore based on a single-level role concept. Table 1, shows the properties of the considered benchmark instances, where ρ_M denotes the density of a matrix M.

Table 1. Key figures of the benchmark instances of RMPlib.

	$|U|$	$|R|$	$|P|$	ρ_{UA}	ρ_{PA}
PLAIN_small_02	50	25	50	0.2	0.1
PLAIN_small_05	100	50	100	0.05	0.05

To investigate the possible influence of the structure of the benchmark instances, two additional benchmark instances were developed for this paper, which are based on two role levels (*TL_01* and *TL_02*). For this purpose, UCA, CSA and SPA matrices were created randomly. The UPA matrix is again obtained from Boolean matrix multiplication $UPA = UCA \otimes CSA \otimes SPA$. The parameters used to create the benchmark instances are based on the parameters of the RMPlib-instances considered and listed in Table 2. To reflect the characteristics of the C2L-RMP, the number of permissions assigned to each single role used for benchmark creation was limited by 5.

Table 2. Parameters used for the creation of new benchmark instances.

	$|U|$	$|C|$	$|S|$	$|P|$	ρ_{UCA}	ρ_{CSA}	ρ_{SPA}
TL_01	100	25	53	100	0.160	0.070	0.035
TL_02	100	25	51	100	0.160	0.078	0.034

7.1 Evaluation of Consecutive Role Mining

In a first evaluation scenario, the influence of s_{max} on the solutions of the C2L-RMP is investigated. Additionally, the two variants of consecutive role mining are compared. For this purpose, three different values of $s_{max} \in \{3, 5, 10\}$ were examined on each of the benchmark instances considered, once using SRF and once using CRF. The parameters of the addRole-EA were adopted from [2]. Each test setup was repeated 20 times with different random seeds for each value of s_{max}. Results were averaged. Fitness values are calculated as total number of composite and single roles $|C| + |S|$. Figure 9 shows the progression of the fitness values of the best individuals per iteration averaged over all values of s_{max}. It can be seen that CRF outperforms SRF in all benchmark instances. This result is also obtained if the different values of s_{max} are considered separately.

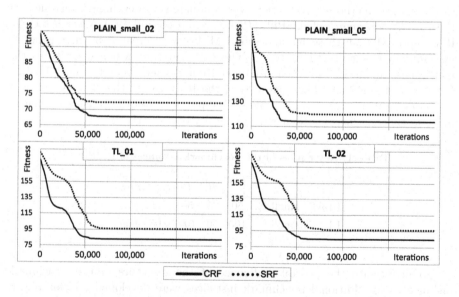

Fig. 9. Comparing solution qualities of CRF and SRF.

Table 3 shows the average occupancy rates of the matrix rows in UCA, CSA and SPA. As CRF can be considered the superior variant of consecutive two-level role mining, evaluation results of SRF are omitted. It is noticeable that, although the number of permissions per single role was limited by 5, creating TL_01 and

TL_02, $|SPA|/|S|$ exceeds this value for $s_{max} = 10$. As could be expected, since if s_{max} was ∞, the C2L-RMP coincides with the B2L-RMP, large values of s_{max} provide sparsely populated CSA matrices, which corresponds to the scenario of Theorem 1, where an optimum solution of the B2L-RMP is obtained, in which only one single role is assigned to each composite role such that $|CSA|/|C| = 1$ and CSA equals the identity matrix.

Table 3. Occupancy rates of matrix rows considering CRF.

Instance	PLAIN_small_02			PLAIN_small_05			TL_01			TL_02		
s_{max}	3	5	10	3	5	10	3	5	10	3	5	10
$|UCA|/|U|$	5.55	6.83	6.09	2.98	2.98	5.75	6.43	13.11	6.07	6.89	6.83	6.91
$|CSA|/|C|$	5.34	2.52	1.60	3.94	1.82	1.14	5.23	3.05	1.68	5.95	3.28	1.77
$|SPA|/|S|$	1.84	3.29	4.99	1.85	3.58	4.91	2.26	3.67	6.78	2.07	3.57	6.57

7.2 Evaluation of Alternating Role Mining

In a second evaluation scenario, the alternating role mining approach is examined more closely. Particular attention is paid to the influence of the number of iterations p for the optimization on each role level before changing to the other role level. Hence, different values of $p \in \{1000, 10000, 20000, 50000\}$ are investigated on each of the considered benchmark instances, using $s_{max} = 5$. The parameters for the addRole-EA were adopted from [2]. Each test setup was repeated 20 times with different random seeds. Results were averaged.

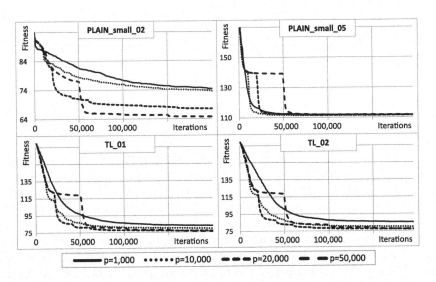

Fig. 10. Comparing solution qualities for different values of p.

It can be seen that higher values of p lead to better results in the long run in almost all cases. However, high values of p seem to delay the optimization of roles especially at the beginning of the optimization process, due to the high number of iterations spent on each role level. This becomes particularly evident for $p = 50,000$. Hence, a possible alternative could be to choose the values of p dynamically and to increase them in the course of the optimization process.

7.3   Comparison of Two-Level Role Mining Approaches

In a final evaluation scenario, the three presented approaches for two-level role mining were compared with each other. For this purpose, the CRF-approach of consecutive role mining, the alternating approach with $p = 20,000$ and the simultaneous role mining approach were each run 20 times with different random seeds on each of the considered benchmark instances, using $s_{max} = 5$. The parameters of the simultaneous approach were also adopted from [2].

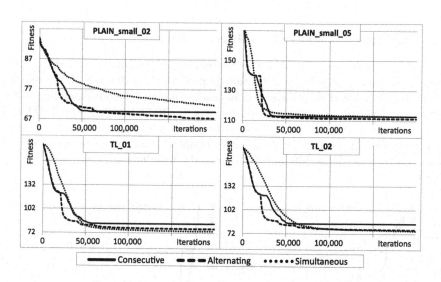

Fig. 11. Comparing solution qualities of two-level role mining approaches.

In contrast to the first two evaluation scenarios, there is a noticeable difference between the two-level benchmark instances provided and the single-level instances of RMPlib. It turns out that the simultaneous approach provides the worst results in terms of the defined fitness $|C| + |S|$ on the instances of RMPlib, while it provides the best results on TL_01 and TL_02. This suggests that the simultaneous approach is particularly strong on data based on a two-level structure. However, it can be seen that the other methods reduce the number of roles much faster at the beginning of the optimization process. In addition, initial experiments suggest that the simultaneous approach is significantly slower than the alternating approach in terms of computation time (between factor 5 and 10 on all instances). Since the consecutive role mining approach involves the stopping condition of the original addRole-EA, a comparison in terms of computing time is rather difficult. A hybrid approach, in which optimization is first performed alternatingly, before switching to the simultaneous approach, would be conceivable in order to increase efficiency.

8 Conclusion and Future Works

In this paper, the Constrained Two-level RMP (C2L-RMP) was defined. In contrast to current approaches in hierarchical role mining, which usually involve many role levels, the C2L-RMP considers a two-level role hierarchy comprising composite and single roles, which allows for application to industry-strength enterprise systems like SAP ERP. In addition, three different approaches to tackle the C2L-RMP were presented. Two of them divide the two-level problem into separate single-level sub-problems, which are optimized individually either consecutively or in an alternating manner. The third approach aims at optimizing both role levels simultaneously. It could be shown that the simultaneous approach provides the best results considering the number of single and composite roles on benchmark instances with two-level role structure. Since the two-level role mining problem is of great practical relevance, it is desirable to apply and examine the developed methods, in real-life business use-cases. Especially in dynamic business environments, the simultaneous approach could be beneficial to address upcoming dynamic events, like employees joining or leaving a company, on composite and single role level at the same time.

References

1. Anderer, S., Kempter, T., Scheuermann, B., Mostaghim, S.: The dynamic role mining problem: role mining in dynamically changing business environments. In: Proceedings of IJCCI 2021, pp. 37–48. INSTICC, SciTePress (2021)
2. Anderer, S., Kreppein, D., Scheuermann, B., Mostaghim, S.: The addRole-EA: a new evolutionary algorithm for the role mining problem. In: Proceedings of IJCCI 2020, pp. 155–166. SciTePress (2020). https://doi.org/10.5220/0010025401550166
3. Anderer, S., Scheuermann, B., Mostaghim, S., Bauerle, P., Beil, M.: RMPlib: a library of benchmarks for the role mining problem. In: Proceedings of SACMAT 2021, SACMAT 2021, pp. 3–13. ACM, New York (2021). https://doi.org/10.1145/3450569.3463566
4. Blundo, C., Cimato, S.: A simple role mining algorithm. In: Proceedings of SAC 2010, pp. 1958–1962. ACM Press, New York (2010). https://doi.org/10.1145/1774088.1774503
5. Dong, L.J., Wang, M.C., Kang, X.J.: Mining least privilege roles by genetic algorithm. Appl. Mech. Mater. **121–126**, 4508–4512 (2011). https://doi.org/10.4028/www.scientific.net/AMM.121-126.4508
6. Du, X., Chang, X.: Performance of AI algorithms for mining meaningful roles. In: 2014 IEEE Congress on Evolutionary Computation (CEC), pp. 2070–2076. IEEE (2014). https://doi.org/10.1109/CEC.2014.6900321

7. Ene, A., Horne, W., Milosavljevic, N., Rao, P., Schreiber, R., Tarjan, R.E.: Fast exact and heuristic methods for role minimization problems. In: Proceedings of SACMAT 2008, pp. 1–10. ACM Press, New York (2008). https://doi.org/10.1145/1377836.1377838

8. Guo, Q., Vaidya, J., Atluri, V.: The role hierarchy mining problem: Discovery of optimal role hierarchies. In: 2008 Annual Computer Security Applications Conference (ACSAC), pp. 237–246. IEEE (2008)

9. Huang, H., Shang, F., Liu, J., Du, H.: Handling least privilege problem and role mining in RBAC. J. Comb. Optim. **30**(1), 63–86 (2013). https://doi.org/10.1007/s10878-013-9633-9

10. Kumar, R., Sural, S., Gupta, A.: Mining RBAC roles under cardinality constraint. In: Jha, S., Mathuria, A. (eds.) ICISS 2010. LNCS, vol. 6503, pp. 171–185. Springer, Heidelberg (2010). https://doi.org/10.1007/978-3-642-17714-9_13

11. Lu, H., Vaidya, J., Atluri, V.: Optimal boolean matrix decomposition: application to role engineering. In: 24th International Conference on Data Engineering, pp. 297–306. IEEE (2008). https://doi.org/10.1109/ICDE.2008.4497438

12. Mitra, B., Sural, S., Vaidya, J., Atluri, V.: A survey of role mining. ACM Comput. Surv. **48**(4), 1–37 (2016). https://doi.org/10.1145/2871148

13. Molloy, I., et al.: Mining roles with semantic meanings. In: Proceedings of SACMAT 2008, pp. 21–30. ACM Press, New York (2008). https://doi.org/10.1145/1377836.1377840

14. Molloy, I., Li, N., Li, T., Mao, Z., Wang, Q., Lobo, J.: Evaluating role mining algorithms. In: Proceedings of SACMAT 2009, pp. 95–104. ACM Press, New York (2009). https://doi.org/10.1145/1542207.1542224

15. Saenko, I., Kotenko, I.: Genetic algorithms for role mining problem. In: PDP 2011, pp. 646–650. IEEE (2011). https://doi.org/10.1109/PDP.2011.63

16. Saenko, I., Kotenko, I.: Reconfiguration of RBAC schemes by genetic algorithms. In: IDC 2016. SCI, vol. 678, pp. 89–98. Springer, Cham (2017). https://doi.org/10.1007/978-3-319-48829-5_9

17. Sandhu, R.S., Coyne, E.J., Feinstein, H.L., Youman, C.E.: Role-based access control models. Computer **29**(2), 38–47 (1996). https://doi.org/10.1109/2.485845

18. Schlegelmilch, J., Steffens, U.: Role mining with ORCA. In: Proceedings of SACMAT 2005, pp. 168–176. ACM Press, New York (2005). https://doi.org/10.1145/1063979.1064008

19. Takabi, H., Joshi, J.B.: Stateminer: an efficient similarity-based approach for optimal mining of role hierarchy. In: Proceedings of the 15th ACM Symposium on Access Control Models and Technologies, pp. 55–64 (2010)

20. Vaidya, J., Atluri, V., Guo, Q.: The role mining problem. In: Proceedings of SACMAT 2007, pp. 175–184. ACM Press, New York (2007). https://doi.org/10.1145/1266840.1266870

21. Vaidya, J., Atluri, V., Warner, J., Guo, Q.: Role engineering via prioritized subset enumeration. IEEE Trans. Dependable Secure Comput. **7**(3), 300–314 (2010). https://doi.org/10.1109/TDSC.2008.61

22. Zhang, D., Ramamohanarao, K., Ebringer, T.: Role engineering using graph optimisation. In: Proceedings of SACMAT 2007, pp. 139–144. ACM Press, New York (2007). https://doi.org/10.1145/1266840.1266862

23. Zhang, D., Ramamohanarao, K., Versteeg, S., Zhang, R.: Graph based strategies to role engineering. In: Proceedings of CSIIRW 2010, pp. 1–4. ACM Press, New York (2010). https://doi.org/10.1145/1852666.1852694

Simplifying Dispatching Rules in Genetic Programming for Dynamic Job Shop Scheduling

Sai Panda, Yi Mei$^{(\boxtimes)}$, and Mengjie Zhang

Victoria University of Wellington, PO Box 600, Kelburn, New Zealand
panda.sai@pm.me, {yi.mei,mengjie.zhang}@ecs.vuw.ac.nz

Abstract. Evolving dispatching rules through Genetic Programming (GP) has been shown to be successful for solving Dynamic Job Shop Scheduling (DJSS). However, the GP-evolved rules are often very large and complex, and are hard to interpret. Simplification is a promising technique that can reduce the size of GP individuals without sacrificing effectiveness. However, GP with simplification has not been studied in the context of evolving DJSS rules. This paper proposes a new GP with simplification for DJSS, and analyses its performance in evolving both effective and simple/small rules. In addition to adopting the generic algebraic simplification operators, we also developed new problem-specific numerical and behavioural simplification operators for DJSS. The results show that the proposed approach can obtain better and simpler rules than the baseline GP and existing GP algorithms with simplification. Further analysis verified the effectiveness of the newly developed numerical and simplification operators.

Keywords: Dynamic Job Shop Scheduling · Genetic Programming · Simplification

1 Introduction

Job Shop Scheduling (JSS) seeks to create a schedule for jobs arriving at a job shop by a set of machines while simultaneously optimising some objective (e.g., trying to finish all jobs before their due date). Typically, the schedule must satisfy the *resource constraint* and *precedence constraint*. The resource constraint indicates that a machine in the job shop can only process at most one job at any given time. The precedence constraint means that an operation of a job cannot be processed until its precedent operation has been completed. JSS has many real-world applications, such as factory production [11] and manufacturing [34].

In real world, unpredicted new job arrivals can occur during the execution of the pre-planned schedule. This is referred to as Dynamic JSS (DJSS). Traditional solution optimisation methods such as mathematical programming [3,23] and meta-heuristics [13,32] are difficult to apply to DJSS, since they cannot adjust the pre-planned schedule efficiently in a dynamic environment.

© The Author(s), under exclusive license to Springer Nature Switzerland AG 2022
L. Pérez Cáceres and S. Verel (Eds.): EvoCOP 2022, LNCS 13222, pp. 95–110, 2022.
https://doi.org/10.1007/978-3-031-04148-8_7

A common method to solve DJSS is to use *dispatching rules* [2,29] to make real-time scheduling decisions in a reactive fashion. Specifically, during the scheduling process, once a machine becomes idle (e.g., completes its current job) and its job queue is not empty, a dispatching rule is used to select the next operation from its queue to process. Commonly used dispatching rules include First-Come-First-Serve (FCFS), Shortest-Processing-Time (SPT) and Earliest-Due-Date (EDD).

The real-world JSS scenarios vary in different aspects such as shop utilisation level (how frequent new jobs arrive and how busy the job shop is) and objectives to be optimised. On the other hand, different dispatching rules are suitable for different JSS scenarios. Thus, it is challenging to design/select an effective dispatching rule for a specific JSS scenario to be solved.

Due to its flexible representation and strong learning ability, Genetic Programming (GP) has been successfully applied as a data-driven approach to learn effective dispatching rules for DJSS [4,25]. GP evolves a population of dispatching rules, and uses a set of training JSS instances (e.g., from the historical data) to evaluate the fitness of the evolved rules.

Although the GP-evolved rules are demonstrated to be much more effective than the manually designed rules, they are usually very large and complex. This can cause serious issues in practice, as the real users cannot understand the evolved rules and thus cannot use them confidently. To address this issue, it is important to improve the GP methods to evolve simpler and easier-to-understand rules.

There have been various existing studies to consider program effectiveness and complexity in GP, such as bloat control and GP with simplification. In this paper, we focus on the program simplification techniques, based on the motivation that the GP individuals usually have a large portion of redundant subtrees, which can be simplified without significantly deteriorating effectiveness. Simplification in GP has been studied in different problems such as regression [41] and classification [42]. However, it has not been studied in the context of evolving scheduling rules.

In this paper, we aim to develop a new GP algorithm with simplification for evolving effective and simple dispatching rules for DJSS. We have the following specific research objectives:

1. Design comprehensive GP tree simplification operators based on various properties including the algebraic principles and specific JSS decision making (behavioural) properties.
2. Develop a new GP algorithm with the new simplification operators for JSS.
3. Verify the effectiveness of the newly proposed GP algorithm on a range of DJSS scenarios and analyse the results in terms of rule quality and size.
4. Analyse the effectiveness of the designed algebraic, numerical and behavioural simplification operators.

The rest of the paper is organised as follows. Section 2 gives the problem description and related work. The newly proposed GP algorithm with simplification is described in Sect. 3. The experimental studies are conducted in Sect. 4. The paper is concluded in Sect. 5.

2   Background

2.1   (Dynamic) Job Shop Scheduling

In Job Shop Scheduling (JSS), there are a set of N jobs that must be processed by a set of M machines. The jth ($j \in \{1, \ldots, N\}$) job comprises of a sequence of L_j operations. The ith ($i \in \{1, \ldots, L_j\}$) operation of the jth job can only be processed by a machine $m_{ij} \in \{1, \ldots, M\}$ and its processing time is p_{ij}. The jth job has an arrival time a_j, a due date d_j and a weight w_j.

JSS is to determine the time to start the processing of each operation, i.e., t_{ij}, $i \in \{1, \ldots, L_j\}$, $j \in \{1, \ldots, N\}$, so that the following constraints are satisfied:

- Each machine in the job shop can only process at most one operation at any given time.
- Each operation cannot start processing until its previous operation has been completed.
- For each job, the first operation cannot start processing until it arrives.

The completion time of the jth job is $C_j = t_{L_j j} + p_{L_j j}$. In this paper, we consider to minimise the following objectives

$$\text{Mean tardiness: } T_{mean} = \frac{1}{N} \sum_{j=1}^{N} \max\{C_j - d_j, 0\}, \qquad (1)$$

$$\text{Maximum tardiness: } T_{max} = \max_{j=1,\ldots,N}\{\max\{C_j - d_j, 0\}\}, \qquad (2)$$

$$\text{Mean weighted tardiness: } MWT = \frac{1}{N} \sum_{j=1}^{N} w_j \max\{C_j - d_j, 0\}, \qquad (3)$$

In DJSS, the information of a job (e.g., p_{ij} and m_{ij}) is not known to the scheduler until it arrives at a_j. In other words, at any time t, the scheduler can only know the information about the jobs with $a_j \leq t$. In this case, one cannot make a complete schedule at $t = 0$, and has to adjust the schedule in real time when new jobs arrive.

2.2   Related Work

Existing Methods for JSS and DJSS. The existing methods for JSS can be briefly categorised into *exact* methods, *(meta-)heuristic* methods and *hyper-heuristic* methods. The exact methods for JSS, such as branch and bound [5] and cutting plane [24], can guarantee optimality. However, due to the NP-hardness of JSS, the effectiveness of the exact methods are restricted to only the small problem instances.

The (meta-)heuristic methods for JSS mainly consists of *construction* heuristics and *iterative improvement* heuristics. The construction heuristics builds a complete solution from scratch, by adding one operation processing into the schedule at a time. The heuristic stops when a complete solution has been built.

Dispatching rules [30] are commonly used construction heuristics. Specifically, once a machine becomes idle and its queue is not empty, a dispatching rule is used to calculate the priority value of each operation in the queue. Then, the candidate operation with the highest priority value is selected to be process next by the idle machine. There have been a number of studies to manually design effective dispatching rules for a range of job shop scenarios (e.g., [10,16,17]). In contrast, the iterative improvement heuristics search in the complete solution space, trying to find better solutions. The representative iterative improvement heuristics include simulated annealing [19], tabu search [9] and genetic algorithms [28]. They can obtain better solutions than the construction heuristics at the cost of higher computational complexity. Therefore, in a dynamic environment where decisions must be made in real time, the construction heuristics such as dispatching rules are more efficient than the iterative improvement heuristics.

The hyper-heuristic approaches [7] search in heuristic space rather than solution space, aiming to find effective heuristics for a family of problem instances. The existing hyper-heuristics for JSS mainly consists of *selection* hyper-heuristics (e.g., [12]) that selects from low-level heuristics to form high-level heuristics and *generative* hyper-heuristics (e.g., [4,25] that generates high-level heuristics from raw features.

Genetic Programming Hyper-heuristics. GP hyper-heuristics have been successfully applied to learn heuristics in different problems [6]. For evolving DJSS dispatching rules, extensive studies have been conducted to address various design issues, such as feature selection [21,39], surrogate evaluations [14,38,40] and solving multiple JSS scenarios simultaneously [27,37].

Apart from JSS, GP has been successfully applied to other problem domains, such as packing [8], vehicle routing [20], timetabling [1] and cloud resource allocation [31]. The main issue is to design a framework for the heuristic to work in (e.g., the dispatching rule in the JSS process). GP is then used to evolve the heuristic, which is typically a priority function represented as a GP tree.

GP with Simplification. It is known that the individuals in the GP population tends to have many redundant branches, which can be removed without affecting the behaviour of the GP tree. It is desirable to remove such redundant branches from the obtained GP trees to improve its interpretability. To this end, several studies have been done to simplify GP trees (e.g., [15,18,35,36]).

For the simplification operators, some studies used the *algebraic* simplification operators (e.g. [35,36], such as simplifying $A - A$ to 0 and A/A to 1. The algebraic operators are directly based on the genotype structures of the GP tree. They are generic and can be applied to all types of problems.

However, the algebraic simplification operators have some limitations that they can hardly identify the indirect redundancies where a smaller tree has a quite different structure from a large tree but has the same behaviour. To address this issue, the *numerical* simplification operators (e.g., [15,18]) have been developed. The main idea of numerical simplification is to estimate the contribution of each node to its parent node. A node (and its sub-tree) is removed if its contribution to its parent node is negligible, and the parent is replaced with the other child branch. Here, the contribution estimation is the key issue. The permutation test [15] and range estimation [18] have been commonly used to estimate the node contributions.

For DJSS, the only existing work for GP with simplification was found in [26]. In this work, the existing generic algebraic simplification operators were adopted, and a number of specific algebraic simplification operators to DJSS that takes the signs of the raw terminals to simplify the trees. For example, $\min\{X+\texttt{ProcTime}, X\}$ can be simplified to X, since $\texttt{ProcTime}$ is always positive. The resultant GP with simplification has been demonstrated to evolve smaller rules without deteriorating effectiveness.

However, it was also observed that the algebraic simplification operators are not so powerful for DJSS, as the number of individuals that can be simplified is limited, especially after the first generation. The numerical simplification operators are expected to be more powerful and can simplify more individuals. However, they are less accurate than the algebraic operators. In addition, the node contribution estimation of the existing numerical simplification operators may not be directly applicable to DJSS, due to the different characteristics of the problem domains.

To overcome the above issue, in this paper, we aim to develop a new GP with comprehensive simplification for DJSS, which includes the *Algebraic* simplification operators as well as newly developed specific *Numeric* and *Behavioural* simplification operators. The proposed algorithm is thus named ANBGP. The details of the proposed ANBGP is described in the next section.

3   Proposed ANBGP

The framework of the proposed ANBGP is shown in Fig. 1. First, a population of GP individuals (each representing a priority rule) are randomly initialised by ramp-half-and-half. At each generation, each individual is first evaluated by the DJSS training instances/simulations $\mathcal{S}^{train}$. Given a dispatching rule x and a DJSS simulation σ, a schedule y can be obtained by running the simulation with the dispatching rule, denoted as $y = \texttt{simrun}(x, \sigma)$. The fitness of a dispatching rule x is calculated based on the objective value (e.g., mean tardiness) of the obtained schedules, i.e.,

$$\texttt{fit}(x) = \frac{1}{|\mathcal{S}^{train}|} \sum_{\sigma \in \mathcal{S}^{train}} f(\texttt{simrun}(x, \sigma)), \tag{4}$$

where $f(\cdot)$ is the objective function of a schedule.

After the evaluation, the individuals are simplified by the behavioural, numerical and algebraic simplifications, whose details will be given in Sect. 3.1. The simplification comes after the evaluation, since the behavioural and numerical simplification require the behaviours of the subtrees obtained from the evaluation process. If the stopping criteria are met, then the best individual in the population is returned. Otherwise, new offspring will be generated by crossover, reproduction and mutation operators.

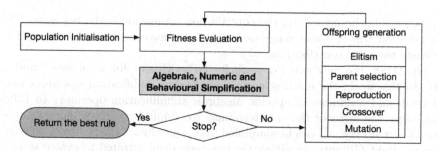

Fig. 1. The framework of the proposed ANBGP.

Compared with existing GP approaches for DJSS, the main difference of ANBGP is the simplification step with the adopted algebraic simplification and newly developed numerical and behavioural simplification.

3.1 Simplification

Algorithm 1 shows the simplification process of a GP tree x. It simply applies the behavioural, numerical and algebraic simplification operators in order.

Algorithm 1: `simplify(x)`

Input: A GP tree x, a set of decision situations Θ.
Output: The simplified tree x'.

1 $x_1 = \texttt{behaviour\_simplify}(x, \Theta)$;
2 $x_2 = \texttt{numerical\_simplify}(x_1, \Theta)$;
3 $x' = \texttt{algebraic\_simplify}(x_2)$;
4 **return** x';

The behavioural simplification operator simplifies a tree with a smaller sub-tree that shows the same behaviour. It is described in Algorithm 2. First, for each decision situation $\theta \in \Theta$, we calculate the behaviours (priority values of the candidate operations) of x and all its subtrees $\tau \in subtrees(x)$. Then, x is simplified with the smallest sub-tree that behaves the same as x in all the decision situations in Θ.

Figure 2 shows an example of the behaviour a PT + WINQ + AT/AT tree (the meaning of the terminals will be explained in Table 2) in a specific DJSS decision situation with three candidate operations. Each subtree is associated with a 3-dimensional behaviour vector, standing for the priority value of the three candidate operations by the subtree. It can be seen that the right branch is AT/AT = 1 and thus redundant. For the left branch, both the subtree PT + WINQ and the terminal PT makes the same decision as the entire tree (selecting the second candidate operation). If Θ contains only this single decision situation, then the simplified tree will be PT, since it has a smaller size than PT.

Algorithm 2: `behaviour_simplify(x, Θ)`

Input: A GP tree x, a set of decision situations Θ.
Output: The simplified tree x'.

1 **foreach** $\tau \in subtrees(x)$ **do** $c(\tau) = 0$;
2 **foreach** *decision situation* $\theta \in \Theta$ **do**
3   Calculate the behaviour of x and the decisions $\{d^\theta(\tau) \mid \tau \in subtrees(x)\}$;
4   **foreach** $\tau \in subtrees(x)$ **do**
5   **if** $d^\theta(\tau) = d^\theta(root(x))$ **then** $c(\tau) = c(\tau) + 1$;
6   **end**
7 **end**
8 $x' = x$;
9 **foreach** $\tau \in subtrees(x)$ **do**
10   **if** $c(\tau) = |\Theta|$ *and* $size(\tau) < size(x')$ **then** $x' = \tau$;
11 **end**
12 **return** x';

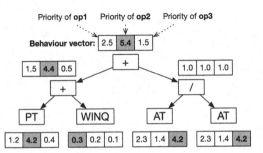

Fig. 2. An example of the behaviour of a PT + WINQ + AT/AT tree.

The newly developed numerical simplification operator is based on the range estimation of the output of the subtrees. The brief idea is similar to [18], which removes a subtree if its output has a significantly smaller range than its parent. The numerical simplification operator is described in Algorithm 3. First, for each decision situation, the range of the outputs of each subtree is calculated for this decision situation. Then, for each subtree, if its range is negligible comparing with the range of its parent (with the threshold coefficient of $\epsilon = 0.1\%$ in line 5), then it can be removed for this decision situation. In the end, we examine all the subtrees that can be removed in all the decision situations (i.e., $c(\tau) = |\Theta|$), and remove the subtrees recursively from the terminals.

Figure 3 shows an example of the range calculation of a $X + Y$ tree in two decision situations, each with three candidate operations. The right-hand side gives the X and Y values of each candidate operation in each decision situation. From this example, we can see that the range of X in DS1 is $0.04 - 0.02 = 0.02$, which is negligible for the range of the whole tree ($802.02 - 2.04 = 799.98$). However, this is not true in DS2. Therefore, the tree cannot be simplified numerically since X cannot be removed in all the decision situations.

Algorithm 3: numerical_simplify(x, Θ)

Input: A GP tree x, a set of decision situations Θ.
Output: The simplified tree x'.

1   **foreach** $\tau \in subtrees(x)$ **do** $c(\tau) = 0$;
2   **foreach** *decision situation* $\theta \in \Theta$ **do**
3 | Calculate $\{range(\tau) \mid \tau \in subtrees(x)\}$;
4 | **foreach** $\tau \in subtrees(x)$ **do**
5 | | **if** $range(\tau) < \epsilon \times range(parent(\tau))$ **then** $c(\tau) = c(\tau) + 1$;
6 | **end**
7   **end**
8    $x' = x$;
9   **foreach** $\tau \in subtrees(x)$ *starting from terminals* **do**
10 | **if** $c(\tau) = |\Theta|$ **then** Replace $parent(\tau)$ with the other branch;
11   **end**
12   **return** x';

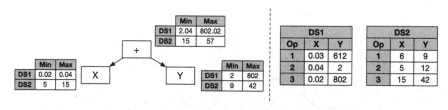

Fig. 3. An example of the range calculation of a $X + Y$ tree in two decision situations.

The algebraic simplifications are directly adopted from [26]. Due to the space limit, the details are not given here, as they can be found from [26].

Selection of Decision Situations. In the newly developed behavioural and numerical simplification operators, the decision situations Θ plays a major role. On one hand, we need a comprehensive set of decision situations for accurate estimation. On the other hand, the extra simplification cost will increase as the increase of the decision situations. To this end, we select the decision situations as follows.

- The decision situations are selected from those encountered during the fitness evaluation. This can avoid extra computation of behaviours.
- We select 100 decision situations with at least 20 candidate operations in the queue, to achieve good balance between computational efficiency and accuracy.

4   Experimental Studies

To verify the effectiveness of the proposed ANBGP, we compare it with the following existing GP methods on a range of different DJSS scenarios.

- BaselineGP [22]: the baseline GP without simplification
- AGP [26]: the GP with only algebraic simplification
- ANGP: the GP with the proposed numerical and algebraic simplification (removing line 1 in Algorithm 1)

In the experiments, we consider the three objectives shown in Eqs. (1)–(3) and two utilisation levels (0.85 and 0.95, indicating that the job shop is busy for 85% and 95% of the time on average), leading to $3 \times 2 = 6$ different DJSS scenarios. Each DJSS scenario is denoted as $\langle obj, ulevel \rangle$. For example, $\langle Tmean, 0.95 \rangle$ stands for the scenario that minimises the mean tardiness and has a utilisation level of 0.95.

For each scenario, we use a set of training instances to evaluate the GP rules during the training process, and use an unseen set of test instances to examine the out-of-bag performance of the best evolved GP rules. Specifically, each training/test instance is a DJSS simulation with 10 machines, 1000 warmup jobs and 5000 jobs after warmup. The job arrivals follow a Poisson process, where the arrival rate is set based on the utilisation level. The number of operations for a job is randomly sampled from 2 to 10, and the processing time of each operation is sampled between 1 and 99.

During the training process, we use one training simulation, and rotate its random seed at each new generation. The test instances include 50 unseen simulations. To avoid the bias caused by the different scales of the test instances, the test performance of a rule x is normalised as follows.

$$perf_{test}(x) = \frac{1}{|\mathcal{S}^{test}|} \sum_{\sigma \in \mathcal{S}^{test}} \frac{f(\texttt{simrum}(x, \sigma))}{f(\texttt{simrum}(x^*, \sigma))}, \tag{5}$$

where x^* is a reference rule, which is set to the ATC rule [33] for T_{mean}, WATC [33] for MWT and EDD for T_{max}.

The GP parameter settings are shown in Table 1. The terminals are described in Table 2. The function set is $\{+, -, *, /, \max, \min, if\}$, where $/$ returns 1 if divided by zero. Each algorithm is run 30 times independently on each scenario.

Table 1. The GP parameter settings.

Parameter	Value
Population Size	1000
Crossover/mutation/reproduction Rates	80%/15%/5%
Number of Elites	10
Maximum Tree Depth	8
Number Of Generations	50

Table 2. The terminals used in the experiments.

Terminal	Description
NIQ	Number Of Operations in the Queue
WIQ	Work in the Queue
MWT	Machine Waiting Time
PT	Processing Time of the Operation
NPT	Processing Time of the next Operation
OWT	Operation Waiting Time
NWT	Next Waiting Time
WKR	Work Remaining
NOR	Number of Operations remaining
WINQ	Work in the next Queue
NINQ	Number of Operations in the next Queue
rFDD	Relative Flow Due Date
rDD	Relative Due Date
W	Job Weight
AT	Arrival Time
SL	Slack

4.1 Test Performance

Figure 4 shows the boxplot of the test performance of the compared algorithms in the six DJSS scenarios. From the figure, one can see that the four compared algorithms show similar test performance, and ANBGP shows slightly better performance on some scenarios (e.g., $\langle T_{max}, 0.85 \rangle$, $\langle MWT, 0.85 \rangle$ and $\langle T_{mean}, 0.95 \rangle$). We have conducted Wilcoxon rank sum test with significance level of 0.05 was also conducted to compare between ANBGP and each of the compared algorithms, and found that ANBGP significantly outperformed all the other algorithms on $\langle T_{max}, 0.85 \rangle$, and showed significantly better test performance than AGP and ANGP on $\langle T_{mean}, 0.95 \rangle$. There was no significant difference on all the other scenarios.

4.2 Rule Size

Figure 5 shows the boxplot of the compared algorithms. The figure shows that ANBGP tend to obtain smaller rule size than the other algorithms on some scenarios. The statistical significance test showed that ANBGP performed significantly better than BaselineGP on three scenarios ($\langle T_{max}, 0.85 \rangle$, $\langle T_{max}, 0.95 \rangle$ and $\langle MWT, 0.95 \rangle$), than AGP on one scenario, and than ANGP on two scenarios. There was no signficant difference in all other scenarios.

To better understand how the proposed simplification operators behave during the GP process, Figure 6 shows the average tree size of all the individuals in the population at each generation of the GP process. From the figure, it can be clearly seen that ANBGP managed to reduce the average tree size in the

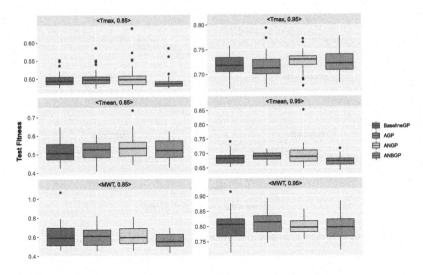

Fig. 4. Test performance boxplots of all algorithms.

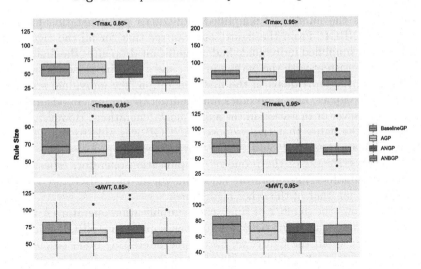

Fig. 5. Best rule size boxplots of all algorithms.

population in $\langle T_{max}, 0.85 \rangle$, $\langle T_{max}, 0.95 \rangle$, $\langle T_{mean}, 0.95 \rangle$ and $\langle MWT, 0.95 \rangle$, where the curves of ANGBP were consistently lower than that of the other compared algorithms. In addition, it can also be seen that the curves of ANGP were lower than AGP on $\langle T_{max}, 0.85 \rangle$, $\langle T_{mean}, 0.95 \rangle$ and $\langle MWT, 0.95 \rangle$. This verifies the effectiveness of the proposed numerical simplification and behavioural simplification operators, as they managed to reduce the size of more individuals in the population.

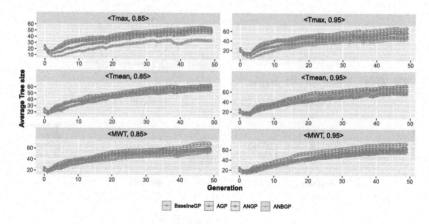

Fig. 6. Average rule size per generation curves of all algorithms.

4.3   Further Analysis

Figure 7 shows the number of simplified individuals per generation for AGP, ANGP and ANBGP. From the figure, one can see that for all the algorithms, the number of simplified individuals had a significant drop after the first generation. This is consistent with the finding in [26], and a possible reason is that after simplifying all the individuals in the first generation, the individual structures have become compact, and it becomes much harder to simplify an offspring generated by crossover or mutation in the subsequent generations. However, we still noticed from the figure that the proposed ANBGP was able to simplify much more individuals than AGP and ANGP during the GP process. This demonstrates the effectiveness of the proposed simplification operator, especially the behavioural simplification operator, since the curves of ANBGP were consistently higher than that of the other algorithms.

Figure 8 shows the number of times each simplification operator (A for algebraic, N for numerical, B for behavioural) was applied in ANBGP in different DJSS scenarios. From the figure, we have some interesting observations. First, it can be seen that the algebraic simplification operator was almost always the most frequently used. Although the numerical and behavioural operators were less used than the algebraic operators, they are much more often used in the busier DJSS scenarios (utilisation level is 0.95), especially the behavioural operators.

Table 3 shows the average training time (in minutes) for each algorithm. It can be seen that all the GP with simplification operators were slower than BaselineGP, due to the extra computational cost of the simplification operators. AGP was the slowest, despite only the algebraic operator is used. This is because that the tree size in AGP are generally larger than ANGP and ANBGP. ANBGP was faster than AGP and ANGP, thanks to the smaller tree size, making the evaluation and simplification more efficient.

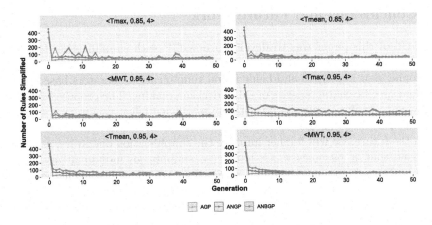

Fig. 7. Number of simplified individuals per generation of all algorithms.

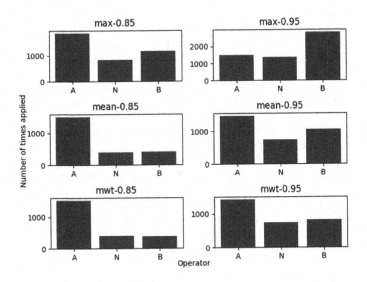

Fig. 8. Average number of times each simplification operator was applied in ANBGP.

Table 3. Average training time over 30 independent runs (in minutes)

Scenario	BaselineGP	AGP	ANGP	ANBGP
$\langle Tmax, 0.85 \rangle$	33.14	44.74	41.50	32.38
$\langle Tmean, 0.85 \rangle$	35.58	43.25	45.01	41.98
$\langle MWT, 0.85 \rangle$	35.41	45.44	44.39	42.50
$\langle Tmax, 0.95 \rangle$	79.41	99.67	113.33	102.49
$\langle Tmean, 0.95 \rangle$	64.32	90.46	86.82	85.20
$\langle MWT, 0.95 \rangle$	71.78	89.83	94.24	96.68

5   Conclusions and Future Work

The goal of this paper was to develop a new GP with simplification to evolve effective and compact dispatching rules for DJSS. This goal has been achieved by developing new numerical and behavioural simplification operators based on the problem specific characteristics of DJSS decision making. The proposed ANBGP managed to evolve smaller rules without sacrificing effectiveness on a range of test set of DJSS scenarios.

In the future, we will further improve the simplification operators, especially the issue that it becomes much harder to simplify an individual again after the initial simplification at the first generation. We will enhance the behaviour simplification operator so that it may be applied to the non-root nodes as well. We will also develop new algorithms to better balance the effectiveness and degree of simplification.

References

1. Bader-El-Den, M., Poli, R., Fatima, S.: Evolving timetabling heuristics using a grammar-based genetic programming hyper-heuristic framework. Memetic Comput. **1**(3), 205–219 (2009)
2. Blackstone, J.H., Phillips, D.T., Hogg, G.L.: A state-of-the-art survey of dispatching rules for manufacturing job shop operations. Int. J. Prod. Res. **20**(1), 27–45 (1982)
3. Blazewicz, J., Dror, M., Weglarz, J.: Mathematical programming formulations for machine scheduling: a survey. Eur. J. Oper. Res. **51**(3), 283–300 (1991)
4. Branke, J., Nguyen, S., Pickardt, C.W., Zhang, M.: Automated design of production scheduling heuristics: a review. IEEE Trans. Evol. Comput. **20**(1), 110–124 (2015)
5. Brucker, P., Jurisch, B., Sievers, B.: A branch and bound algorithm for the job-shop scheduling problem. Discret. Appl. Math. **49**(1–3), 107–127 (1994)
6. Burke, E.K., Hyde, M.R., Kendall, G., Ochoa, G., Ozcan, E., Woodward, J.R.: Exploring hyper-heuristic methodologies with genetic programming. In: Mumford, C.L., Jain, L.C. (eds.) Computational Intelligence, pp. 177–201. Springer, Heidelberg (2009). https://doi.org/10.1007/978-3-642-01799-5_6
7. Burke, E.K., Hyde, M., Kendall, G., Ochoa, G., Özcan, E., Woodward, J.R.: A classification of hyper-heuristic approaches. In: Gendreau, M., Potvin, J.Y. (eds.) Handbook of Metaheuristics, pp. 449–468. Springer, Boston (2010). https://doi.org/10.1007/978-1-4419-1665-5_15
8. Burke, E.K., Hyde, M., Kendall, G., Woodward, J.: A genetic programming hyper-heuristic approach for evolving 2-D strip packing heuristics. IEEE Trans. Evol. Comput. **14**(6), 942–958 (2010)
9. Dell'Amico, M., Trubian, M.: Applying tabu search to the job-shop scheduling problem. Ann. Oper. Res. **41**(3), 231–252 (1993)
10. Dominic, P.D., Kaliyamoorthy, S., Kumar, M.S.: Efficient dispatching rules for dynamic job shop scheduling. Int. J. Adv. Manuf. Technol. **24**(1), 70–75 (2004)
11. Gao, D., Wang, G.G., Pedrycz, W.: Solving fuzzy job-shop scheduling problem using de algorithm improved by a selection mechanism. IEEE Trans. Fuzzy Syst. **28**(12), 3265–3275 (2020)

12. Garza-Santisteban, F., et al.: A simulated annealing hyper-heuristic for job shop scheduling problems. In: 2019 IEEE Congress on Evolutionary Computation (CEC), pp. 57–64. IEEE (2019)
13. Gonçalves, J.F., de Magalhães Mendes, J.J., Resende, M.G.: A hybrid genetic algorithm for the job shop scheduling problem. Eur. J. Oper. Res. **167**(1), 77–95 (2005)
14. Hildebrandt, T., Branke, J.: On using surrogates with genetic programming. Evol. Comput. **23**(3), 343–367 (2015)
15. Javed, N., Gobet, F.: On-the-fly simplification of genetic programming models. In: Proceedings of the 36th Annual ACM Symposium on Applied Computing, pp. 464–471. Association for Computing Machinery, New York, March 2021
16. Jayamohan, M., Rajendran, C.: Development and analysis of cost-based dispatching rules for job shop scheduling. Eur. J. Oper. Res. **157**(2), 307–321 (2004)
17. Kaban, A., Othman, Z., Rohmah, D.: Comparison of dispatching rules in job-shop scheduling problem using simulation: a case study. Int. J. Simul. Model. **11**(3), 129–140 (2012)
18. Kinzett, D., Johnston, M., Zhang, M.: Numerical simplification for bloat control and analysis of building blocks in genetic programming. Evol. Intel. **2**(4), 151 (2009)
19. Kolonko, M.: Some new results on simulated annealing applied to the job shop scheduling problem. Eur. J. Oper. Res. **113**(1), 123–136 (1999)
20. Liu, Y., Mei, Y., Zhang, M., Zhang, Z.: Automated heuristic design using genetic programming hyper-heuristic for uncertain capacitated arc routing problem. In: Proceedings of the Genetic and Evolutionary Computation Conference, pp. 290–297 (2017)
21. Mei, Y., Nguyen, S., Xue, B., Zhang, M.: An efficient feature selection algorithm for evolving job shop scheduling rules with genetic programming. IEEE Trans. Emerg. Top. Comput. Intell. **1**(5), 339–353 (2017)
22. Mei, Y., Nguyen, S., Zhang, M.: Evolving time-invariant dispatching rules in job shop scheduling with genetic programming. In: McDermott, J., Castelli, M., Sekanina, L., Haasdijk, E., García-Sánchez, P. (eds.) EuroGP 2017. LNCS, vol. 10196, pp. 147–163. Springer, Cham (2017). https://doi.org/10.1007/978-3-319-55696-3_10
23. Miller, H.E., Pierskalla, W.P., Rath, G.J.: Nurse scheduling using mathematical programming. Oper. Res. **24**(5), 857–870 (1976)
24. Mokotoff, E., Chrétienne, P.: A cutting plane algorithm for the unrelated parallel machine scheduling problem. Eur. J. Oper. Res. **141**(3), 515–525 (2002)
25. Nguyen, S., Mei, Y., Zhang, M.: Genetic programming for production scheduling: a survey with a unified framework. Complex Intell. Syst. **3**(1), 41–66 (2017). https://doi.org/10.1007/s40747-017-0036-x
26. Panda, S., Mei, Y.: Genetic programming with algebraic simplification for dynamic job shop scheduling. In: 2021 IEEE Congress on Evolutionary Computation (CEC), pp. 1848–1855. IEEE (2021)
27. Park, J., Mei, Y., Nguyen, S., Chen, G., Zhang, M.: Evolutionary multitask optimisation for dynamic job shop scheduling using niched genetic programming. In: Mitrovic, T., Xue, B., Li, X. (eds.) AI 2018. LNCS (LNAI), vol. 11320, pp. 739–751. Springer, Cham (2018). https://doi.org/10.1007/978-3-030-03991-2_66
28. Pezzella, F., Morganti, G., Ciaschetti, G.: A genetic algorithm for the flexible job-shop scheduling problem. Comput. Oper. Res. **35**(10), 3202–3212 (2008)
29. Rajendran, C., Holthaus, O.: A comparative study of dispatching rules in dynamic flowshops and jobshops. Eur. J. Oper. Res. **116**(1), 156–170 (1999)

30. Sels, V., Gheysen, N., Vanhoucke, M.: A comparison of priority rules for the job shop scheduling problem under different flow time-and tardiness-related objective functions. Int. J. Prod. Res. **50**(15), 4255–4270 (2012)
31. Tan, B., Ma, H., Mei, Y., Zhang, M.: A cooperative coevolution genetic programming hyper-heuristic approach for on-line resource allocation in container-based clouds. IEEE Trans. Cloud Comput. (2020)
32. Van Laarhoven, P.J., Aarts, E.H., Lenstra, J.K.: Job shop scheduling by simulated annealing. Oper. Res. **40**(1), 113–125 (1992)
33. Vepsalainen, A.P., Morton, T.E.: Priority rules for job shops with weighted tardiness costs. Manag. Sci. **33**(8), 1035–1047 (1987)
34. Wang, H., Jiang, Z., Wang, Y., Zhang, H., Wang, Y.: A two-stage optimization method for energy-saving flexible job-shop scheduling based on energy dynamic characterization. J. Clean. Prod. **188**, 575–588 (2018)
35. Wong, P., Zhang, M.: Algebraic simplification of GP programs during evolution. In: Proceedings of the 8th Annual Conference on Genetic and Evolutionary Computation, pp. 927–934 (2006)
36. Wong, P., Zhang, M.: Effects of program simplification on simple building blocks in genetic programming. In: 2007 IEEE Congress on Evolutionary Computation, pp. 1570–1577. IEEE (2007)
37. Zhang, F., Mei, Y., Nguyen, S., Tan, K.C., Zhang, M.: Multitask genetic programming based generative hyper-heuristics: a case study in dynamic scheduling. IEEE Trans. Cybern. (2021)
38. Zhang, F., Mei, Y., Nguyen, S., Zhang, M.: Collaborative multifidelity-based surrogate models for genetic programming in dynamic flexible job shop scheduling. IEEE Trans. Cybern. (2021)
39. Zhang, F., Mei, Y., Nguyen, S., Zhang, M.: Evolving scheduling heuristics via genetic programming with feature selection in dynamic flexible job shop scheduling. IEEE Trans. Cybern. **51**(4), 1797–1811 (2021)
40. Zhang, F., Mei, Y., Nguyen, S., Zhang, M., Tan, K.C.: Surrogate-assisted evolutionary multitasking genetic programming for dynamic flexible job shop scheduling. IEEE Trans. Evol. Comput. **25**(4), 651–665 (2020)
41. Zhang, M., Wong, P.: Explicitly simplifying evolved genetic programs during evolution. Int. J. Comput. Intell. Appl. **7**(02), 201–232 (2008)
42. Zhang, M., Wong, P.: Genetic programming for medical classification: a program simplification approach. Genet. Program Evolvable Mach. **9**(3), 229–255 (2008)

Novelty-Driven Binary Particle Swarm Optimisation for Truss Optimisation Problems

Hirad Assimi[1]([⊠]) [iD], Frank Neumann[1] [iD], Markus Wagner[1] [iD], and Xiaodong Li[2] [iD]

[1] University of Adelaide, Adelaide, Australia
{hirad.assimi,frank.neumann,markus.wagner}@adelaide.edu.au
[2] RMIT University, Melbourne, Australia
xiaodong.li@rmit.edu.au

Abstract. Topology optimisation of trusses can be formulated as a combinatorial and multi-modal problem in which locating distinct optimal designs allows practitioners to choose the best design based on their preferences. Bilevel optimisation has been successfully applied to truss optimisation to consider topology and sizing in upper and lower levels, respectively. We introduce exact enumeration to rigorously analyse the topology search space and remove randomness for small problems. We also propose novelty-driven binary particle swarm optimisation for bigger problems to discover new designs at the upper level by maximising novelty. For the lower level, we employ a reliable evolutionary optimiser to tackle the layout configuration aspect of the problem. We consider truss optimisation problem instances where designers need to select the size of bars from a discrete set with respect to practice code constraints. Our experimental investigations show that our approach outperforms the current state-of-the-art methods and it obtains multiple high-quality solutions.

1 Introduction

Trusses are used in the construction of bridges, towers [12,30], aerospace structures [31] or in robots [10] and they carry applied external forces on nodes to support structures. Truss topology optimisation is essentially a combinatorial optimisation problem. Topology optimisation aims to decide whether to include or exclude necessary bars so that the structure's weight is as light as possible while adhering to structural constraints.

The ground structure method [34] gives the excessive potential connections between nodes and the preliminary nodal positions in the design space, representing an upper bound in the combinatorial search space of the topology. Therefore, the optimal design is found as a subset of ground structure. Size and shape optimisation are also aspects of truss optimisation, where size optimisation determines the optimal size of active bars and shape optimisation determines the optimal position of nodes in the truss layout.

© The Author(s), under exclusive license to Springer Nature Switzerland AG 2022
L. Pérez Cáceres and S. Verel (Eds.): EvoCOP 2022, LNCS 13222, pp. 111–126, 2022.
https://doi.org/10.1007/978-3-031-04148-8_8

Achieving an overall minimum weight of a truss is the most common objective to save costs and improve efficiency [3] in truss optimisation problems based on ground structure. Truss optimisation is challenging due to its nature, and it has been an active research area for decades in structural engineering. It is important because it can be used to quickly find preliminary designs for further detailed investigation and design [33].

Bilevel optimisation [32] has been a practical approach for tackling truss optimisation problems while dealing with different design variables and considering interactions among them [1,17]. Topology optimisation trusses is a combinatorial and multi-modal problem that for bilevel optimisation of trusses, we consider the topology search space in the upper level and size and shape optimisation of the truss in the lower level.

1.1 Related Work

Truss optimisation problems are subject to structural constraints such as stability, failure criteria and design codes by practice and manufacturing specifications. The constraints can conflict with objectives which makes the problem more challenging.

For decades, several numerical methods have been applied to different truss optimisation problems. Conventional methods such as gradient-based showed limited efficiency in dealing with structural constraints and handling the discreteness and discontinuities in truss optimisation problems which led to trapping in local optima [5]. The inadequacy of classical optimisation methods led to the development of population-based algorithms and metaheuristics in truss optimisation [21]. According to the literature, two-stage approaches for topology, size and shape optimisation consider these design variables linearly separable and find the optimal topology of all active bars of equal size. Next, it determines the optimal sizing for the obtained topology. The assumption of linear separability can lead to missing out good solutions and may result in infeasible solutions with respect to real practice [5]. Other approaches, namely single-stage approaches, consider topology, size and shape design variables simultaneously and consequently expand the search space considering both design variables. The search is guided towards finding one ultimate solution and still may ignore interactions among different variables.

Bilevel optimisation is a more effective approach than the previous two approaches to dealing with truss optimisation problems because it can explicitly model the interaction among different aspects of the problem. In the bilevel formulation, the upper level optimisation problem determines the truss configuration, such as topology, where the lower level optimises bars' sizing. Islam et al. [17] adopted a bilevel representation for the truss optimisation problem, where the weight of the truss was the main optimisation objective for both upper and lower levels. In the upper level, a modified binary Particle Swarm Optimisation (PSO) with niching capability was used to enhance its population diversity while still maintaining some level of exploitation. The niching technique was based on a speciation scheme that applies a niching radius to subdivide the swarm to locate multiple high-quality solutions. A standard continuous PSO was employed for the lower level to supply good sizing solutions to the upper level.

However, using standard PSO for such constrained engineering problems can lead to trapping in local optima [15]. It has been shown that truss optimisation should incorporate domain-specific information of the problem instead of only considering pure optimisation [2]. It has been observed that there exist multiple distinct topologies with almost equal overall weight in the truss optimisation search space [5,17,26]. Therefore finding multiple equally good truss designs with respect to the topology, size and shape can enable practitioners to choose according to their preferences.

Niching methods [26] and novelty search [23] are capable of finding multiple optima in multi-modal search space. Niching techniques employ rules to keep the population's diversity by focusing on optimising a fitness function. However, novelty search drives the search towards different candidate solutions from the ones previously have been encountered. Novelty search was proposed in [23] to improve the exploration capability in population-based algorithms while still maintaining diversity by looking for more novel solutions considering their similarity to other previously visited solutions. Novelty search aims to improve diversity with respect to the behavioural space instead of exclusively considering objective function value [24,27].

1.2 Our Contribution

In this paper, we consider the bilevel optimisation of trusses with a primary focus on the upper level as a combinatorial optimisation problem. We propose two approaches considering the upper level search space in the truss test problems. We introduce an exact enumeration approach for rigorous analysis of the upper level search space for small test problems. Exact enumeration iterates over all possible upper level topologies, and we apply the lower level optimisation to every feasible upper level design. Using exact enumeration enables us to remove randomness in the upper level, characterise its search space, and report on the quality of potential designs.

We propose a novelty-driven binary particle swarm optimisation for bilevel truss optimisation for larger problems. We aim to discover new designs at the upper level by maximising novelty and applying lower-level optimisation to obtained novel solutions. Using novelty search can guide the search in the upper level towards unseen topologies instead of using only the overall weight to explore the search space. Therefore we set different objective functions for the upper and lower levels. Our proposed novelty search driven binary PSO for bilevel truss optimisation comprises of a modified binary PSO in the upper level to deal with the topology search space exploration and maximisation of novelty. Note that the upper level of truss optimisation is subject to primary essential constraints. We employ a repair mechanism to fix infeasible produced solutions. Because lower level sizing in truss optimisation is difficult due to the nature of the problem, we employ a reliable evolutionary optimiser that uses domain-based knowledge to determine the size of bars.

We carry out the proposed approaches to truss optimisation test problems. Our experimental investigation shows that our approaches can outperform the current state of art methods and achieve multiple high-quality truss designs. The source code is available at https://github.com/hiraaaad/BinaryNoveltyPSOTruss.

The remainder of the paper is structured as follows. In the following section, we state the bilevel truss optimisation problem. Following that, we introduce the exact enumeration and propose the bilevel novelty search framework which includes the upper level repair operator. We carry out experimental investigations and report on the quality of solutions for different truss test problems. We conclude with some closing thoughts and ideas for further work.

2   Bilevel Truss Optimisation Problem

In this section, we define the bilevel truss problem according to [17]. Next, we explain the lower level optimiser, and afterwards, we provide some preliminaries on binary PSO used in [17] and this study.

2.1   Problem Definition

Bilevel truss optimisation problem embeds an upper level topology optimisation problem into a lower level size and shape optimisation problem. The bilevel optimisation problem can be stated as,

$$
\begin{aligned}
\text{find} \quad & \vec{x}, \vec{y}, \quad \vec{x} \in \{0, 1\}^m \\
\text{optimise} \quad & F(\vec{x}, \vec{y}) \\
\text{subject to} \quad & G_1(\vec{x}), G_2(\vec{x}), G_3(\vec{x}) \\
\text{where} \quad & G_1(\vec{x}) = \text{True} \iff \text{Necessary nodes are active in truss} \\
& G_2(\vec{x}) = \text{True} \iff \text{Truss is externally stable} \\
& G_3(\vec{x}) = \vec{y} \in \text{argmin}\{W(\vec{x}, \vec{y}), g_j(\vec{x}, \vec{y}) \leq 0, j = 1, 2, 3\}
\end{aligned}
$$

where $\vec{x}$ refers to the binary topology variable in the upper level where it shows if a truss bar is active (1) or excluded (0). m shows the length of upper level topology design representation. For instance, in 25-bar truss m is 8 due to symmetry that the 25 members are divided into 8 groups. For the same problem, we can show the upper bound of topology as the ground structure where all bars are active as $\vec{x} = [11111111]$.

$\vec{y}$ denotes the design variable in the lower level optimisation problem including size and shape with respect to the test problem. The size variables of $\vec{y}$ should be selected from an available size set (S). $F(\vec{x}, \vec{y})$ shows the objective function considered in the upper level such as weight minimisation used in [17] or maximising novelty in this study.

Solutions in the upper level should satisfy the topology constraints for feasibility. G_1 enforces that the design should have active nodes that support the truss and carry the external load, because they are necessary elements in the design space's predefined boundary conditions. For example for 10-bar truss

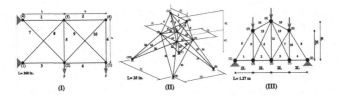

Fig. 1. Ground structure of 10-bar truss (I), 25 bar truss (II) and 15-bar truss (III).

(depicted in Fig. 1 (I)) nodes 1 and 4 are support nodes and nodes 2 and 3 are carrying external loads. Therefore these four nodes are necessary nodes in the design space. G_2 states the external stability of a truss. The external stability satisfaction criteria are fully detailed in Sect. 4.2. Feasible topology solutions should meet G_1 and G_2. In this case, the lower level optimiser aims to find the optimum $\vec{y}$ to minimise the overall weight of truss (W) which is the summation of the weight of all included bars $(\hat{m})$ in the truss.

$$W(\vec{x}, \vec{y}) = \rho \sum_{i=1}^{\hat{m}} x_i y_i l_i$$

where ρ and l show the density of the material used in the truss (such as aluminium or steel) and length of a bar with respect to its end points in the design space, respectively. upper level external stability satisfaction is necessary but not sufficient. Therefore, in the lower level, the internal stability should be checked through lower level function evaluation.

If $(\vec{x}, \vec{y})$ meets the internal stable condition, extra constraints should be satisfied. These constraints state that the computed stress in bars $(\sigma_i, i \in \{1, 2, .., \hat{m}\})$ and displacement of truss nodes $(\delta_k, k = \{1, 2, .., n\})$ after applying the external loads should not exceed their problem-dependent allowable values $\sigma_i^{max}, i \in \{1, 2, .., \hat{m}\}$, and $\delta_k^{max}, k \in \{1, 2, .., n\}$, respectively.

The lower level constraints $g_j, j = 1, 2, 3$, used as part of $G_3(\vec{x})$ are therefore defined as follows.

$$g_1(\vec{x}, \vec{y}) = \text{True} \iff \text{Truss is internally stable}$$
$$g_2(\vec{x}, \vec{y}) \leq 0 \iff g_{2,i}(\vec{x}, \vec{y}) = \sigma_i - \sigma_i^{max} \leq 0 \quad \forall i \in \{1, 2, .., \hat{m}\}$$
$$g_3(\vec{x}, \vec{y}) \leq 0 \iff g_{3,k}(\vec{x}, \vec{y}) = \delta_k - \delta_k^{max} \leq 0 \quad \forall k \in \{1, 2, .., n\}$$

2.2   Lower Level Optimisation

We use a domain knowledge-based evolutionary optimiser for lower level optimisation [2] to determine the optimum layout. The *loweroptimiser* is a variant of Covariance Matrix Adaptation Evolution Strategy algorithm (CMA-ES) that is customised to be problem-specific. *loweroptimiser* follows the main principles of evolutionary strategies to evolve the solutions. However, it uses specific operators to adjust solutions with respect to the allowable stress and displacement in the truss. *loweroptimiser* uses a probabilistic scheme to round the values to the discrete set to avoid biasing the search towards sub-optimal solutions.

loweroptimiser employs a mapping strategy with respect to the response after performing finite element analysis to adjust the sizing of a violating bar by multiplying its current size with a factor that depends on the amount of violation. Another operator is a resizing strategy for producing new individuals near boundary designs of the problem and the problem-dependent constraints. For brevity, we refer the reader to [1,2] for detailed explanations on different components of the *loweroptimiser*.

2.3 Particle Swarm Optimisation (PSO)

PSO is a population-based algorithm evolving a swarm where it contains *particles* as candidate solutions to a problem [19]. Each particle has three vectors at the time (t) of evolution: position $(\vec{z_t})$, velocity $(\vec{v_t})$ and personal best where it keeps the best position it has been evolved to as $(\vec{p_t})$. PSO updates particle positions based on the velocity for each component (i) as follows.

$$\vec{v}_{t+1} = \omega\vec{v_t} + c_1 r_1 \times (\vec{p_t} - \vec{z_t}) + c_2 r_2 \times (\vec{p_g} - \vec{z_t})$$
$$\vec{z}_{t+1} = \vec{z_t} + \vec{v}_{t+1}$$

where $\vec{p_g}$ is the global best position in the swarm. ω is the inertia factor to control the impact of current velocity in velocity updating. r_1 and r_2 are vectors with size of a particle containing random values from a uniform distribution in the range $[0, 1]$. c_1 and c_2 are cognitive and social factors to attract particles toward their personal best and global best.

2.4 Binary PSO

PSO is typically used as a continuous optimisation algorithm. Therefore, to use it for binary search spaces, we need to employ a transfer function, such as Sigmoid transfer function, to map a continuous search space into a binary search space. In this study, we use global topology for the PSO and employ a time-varying transfer function [17] to balance between exploration and exploitation. Velocities of all particles are updated according to the following velocity update equation to determine the probabilities for flipping the position vector elements (i).

$$z_i^t = \begin{cases} 1 & \text{if } rand() \geq TV(v_i^t, \varphi) \\ 0 & \text{otherwise,} \end{cases}$$

where $rand()$ denotes a random value in a uniform distribution in the range $[0, 1]$ and, TV is given as [17],

$$TV(v_i^t \varphi) = \frac{1}{1 + e^{-\frac{v_i^t}{\varphi}}}$$

φ is the control parameter to balance exploration and exploitation in the course of evolution where it linearly decreases from 5.0 to 1.0 in this study according to [17].

Algorithm 1: Exact Enumeration

---

for $i = 1$; $i \leq 2^m$; $i = i + 1$ **do**

    compute $\vec{x}_i$ ▷ `generate the bit string`

    if $\vec{x}_i$ *is feasible* **then**

       | $\vec{y}_i = loweroptimiser(\vec{x}_i)$

    Store $W(\vec{x}_i, \vec{y}_i)$

---

3 Exact Enumeration

We apply exact enumeration to the truss problems where the upper level dimension of the search space is small (m $\leq$ 12). Exact enumeration enables us to enumerate over all possible combinations of binary strings in the search space in the upper level, where each represents a topology design. Therefore we can remove the randomness for these problems from the upper level and investigate its search space rigorously. Algorithm 1 shows the pseudocode of our exact enumeration. This algorithm takes the upper level dimension m as the input and iterates over all possible binary string combinations of m-bits.

If a binary string satisfies the upper level's feasibility criteria G_1 and G_2, it will be sent to the lower level. *loweroptimiser* aims to find the optimum vector for size (and shape), and we store its overall weight.

4 Bilevel Novelty Search

In this section we first introduce the components of our bilevel method, and then we combine these components and introduce the framework of proposed novelty PSO for bilevel truss optimisation.

4.1 Novelty-Driven PSO

Novelty-Driven PSO (NdPSO) is a variant of PSO employing novelty search to drive particles toward novel solutions that are different from previously encountered ones [11]. The main idea is to explore the search space by ignoring objective-based fitness functions and reward novel individuals. NdPSO uses the score of novelty to evaluate the performance of particles. For this purpose, it maintains an archive of past visited solutions to avoid repeatedly cycling through the same series of behaviours. NdPSO evaluates the novelty of particles by computing the average distance of a behaviour to its k-nearest neighbours in the archive as follows:

$$\text{nov}(x) = \frac{1}{k} \sum_{i=0}^{k} dist(x, \mu_i).$$

where μ_i is the ith nearest neighbour of x and *dist* is the Hamming distance metric. Novelty score ensures that individuals in less dense areas with respect to the archive get higher novelty scores.

NdPSO employs core principles of PSO and mainly replaces the objective function with novelty evaluation. Note that personal best and global best value in NdPSO show a dynamic behaviour. For more details on NdPSO, we refer the reader to [11]. We use NdPSO in the upper level of truss optimisation to discover novel topology designs.

4.2   Repair Mechanism in the Upper Level

Topology designs in the upper level are feasible if they meet $G_1(\vec{x})$ and $G_2(\vec{x})$. The following conditions should be satisfied for $G_2(\vec{x})$ [2]:

- The degree of freedom (DoF) in a truss should be non-positive [5].
- The summation of the number of members and restrain forces on a node must be equal or greater than the truss dimension.
- The summation of the number of members and restraint forces on a non-carrying node must be greater than the truss dimension.

As stated in [5], necessary node constraints are more important than the DoF constraint. To deal with infeasible topologies, we use the (1+1)-EA [8] with the following comparator to repair solutions:

$$x \succeq y := (\alpha(x) \leq \alpha(y)) \vee$$
$$(\alpha(x) = \alpha(y) \wedge \beta(x) \leq \beta(y)) \vee$$
$$(\alpha(x) = \alpha(y) \wedge \beta(x) = \beta(y) \wedge \theta(x) \leq \theta(y))$$

where α is the violation degree of active necessary nodes, β is the violation degree of truss DoF and θ is the violation degree of second and third criteria in external stability. (1+1)-EA is a simple evolutionary algorithm where it produces an offspring by mutation and the offspring replaces the parent if the offspring is better with respect to the objective function mentioned above.

4.3   Bilevel Novelty-Driven Binary PSO Framework

Our proposed approach works as follows (see Algorithm 2). Initially, the binary PSO generates a random population of binary strings. Next, the repair mechanism performs on the population to ensure the feasibility of particles. The particles' velocities are drawn randomly from $[-v, v]$. Then, the novelty score is computed for particles with respect to the archive. Because all particles are feasible, *loweroptimiser* computes the corresponding optimal size (and shape) for the upper level topology. With this, we update the archive with the current population. Next, the position and velocity of particles are updated, and the above process repeats till the termination criterion is met.

Algorithm 2: Novelty-Driven Binary PSO for Bilevel Truss Optimisation

---

Randomly generate the initial population of Binary PSO
Repair the initial population into feasible topologies
Set the velocity of particles in population
Evaluate the novelty score for each particle
$\vec{y}_i = loweroptimiser(\vec{x}_i)$
Store $W(\vec{x}_i, \vec{y}_i)$
Update p_t and p_g
Update the archive
repeat
    for *i=1 to population size* **do**
        Update position of particle
        Update velocity of particle
        Repair the particle into feasible upper level solution
        Evaluate novelty score of the particle
        $\vec{y}_i = loweroptimiser(\vec{x}_i)$
        Store $W(\vec{x}_i, \vec{y}_i)$
        Update p_t^i and p_g according to novelty score
        Update the archive
until *termination criterion is met*

---

5   Experimental Investigations

In this study, we use multiple truss test problems with discrete sizing from the literature [2,7,25]. We investigate them in ascending order of length of topology design variable. We use the best reported weights from state-of-art for our comparison.

We split the problems into small and large instances. We apply exact enumeration to small problems where their topology search space is tractable ($m \leq 12$). To show its outcome, we set the ground structure of the problem as an upper bound reference. We sort the other designs with respect to their Hamming distance with this reference design (denoted by d_H). We report on the quality of solutions using the median of best solutions obtained in 30 independent runs in the lower level.

For large instances, we apply the proposed bilevel novelty search PSO and report on the quality of top best-found solutions. We investigate the obtained designs and identification of redundant bars and nodes in the design space. To setup our algorithms, we use the following parameters. For lower level optimisation, we use the parameter settings in [1,2]. For the upper level optimisation, the swarm consisted of 30 particles, $v = 6$, $c_1 = c_2 = 1.0$, ω linearly decreases from 0.9 to 0.4 [17], and the maximum number of iterations is set to 300. We use $k = 3$ nearest neighbours to calculate novelty at the upper level.

Table 1. Comparison of optimised designs for 25-bar truss.

Case 1	[29]	[16], [4], [7]	This Study		
			(a)	(b)	(c)
Best weight (lb)	546.01	484.85	482.6	483.35	484.3

Case 2	[22]	[25], [7]	This Study		
			(a)	(b)	(c)
Best weight (lb)	556.43	551.14	546.97	547.81	548.64

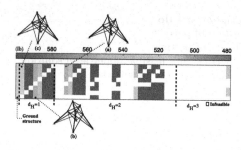

Fig. 2. Exact enumeration on 25-bar truss case 1 (right side truncated). d_H denotes the hamming distance with the upper bound reference. Note that empty area denotes the infeasible region of the search space.

5.1   25-Bar Truss

Figure 1 (II) shows the ground structure of 25-bar truss which is a spatial truss for size and topology optimisation [7]. This is a symmetric truss where 25 truss bars are grouped into eight groups of bars. Therefore, there are 256 possible topologies at the upper level. Figure 2 shows the outcome of exact enumeration on this test problem where we observed that up to first 50 and 55 designs with respect to d_H are feasible and the rest of the search space is infeasible. We can see that most of the feasible designs are in the upper bound vicinity where $d_H \leq 3$. We can also observe that the high-quality solutions exist in the region where $d_H \leq 2$ and $d_H \leq 3$ for cases 1 and 2, respectively. Note that the squares are coloured according to the weight of the obtained design. Table 1 shows our findings for both cases.

5.2   10-Bar Truss

10-bar truss [25] is a well-known size and topology optimisation problem which its ground structure is depicted in Fig. 1 (I). It undergoes single load and the sizing of bars should be selected from a discrete set where there are 10 bars in the topology design, which results in 1024 possible upper level topologies. Figure 3 shows the outcome of exact enumeration and the designs are ordered according to their d_H. We only show the first 320 sorted designs because the rest of the designs are infeasible. This figure also shows the top obtained designs for

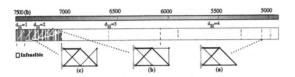

Fig. 3. Exact enumeration on 10-bar truss. d_H denotes the hamming distance with the upper bound reference.

Table 2. Comparison of optimised designs for 10-bar truss.

	[25]	[4], [16], [7]	[9]	[20]	This Study		
					(a)	(b)	(c)
Best weight (lb)	5531.98	5490.74		5056.88	4980.10	4965.70	5107.50	5131.70

this problem. We can see feasible designs are where $d_H \leq 4$ and the best-found design's d_H is 4. Table 2 shows our findings, and we can see that designs (b) and (c) with $d_H = 2$, both identify two bars as redundant and incorporate all six nodes in their designs. However, design (a) which is the best-found design identifies four bars (A_2, $A_5 - A_6$ and A_{10}) as redundant and eliminates node 6. We can also see that other state of art methods, including [9] and [20] also identified the same topology as the optimal topology. However, our approach can obtain a solution with a lower weight due to the efficient lower level optimiser.

5.3   52-Bar Truss

This is a size and topology optimisation problem where 52-bar truss are grouped into 12 bar groups resulting in 4096 possible designs. The truss undergoes three load cases and the sizing of bars should be selected from a discrete set [35]. Figure 4 shows the outcome of exact enumeration where we only show the first 250 designs sorted according to their d_H. Out of all sorted combinations, 1900 designs are feasible, and the rest of the search space is infeasible. We also observed that feasible designs are located where $d_H \leq 6$. Table 3 shows our findings and, we can see that the top three designs for this problem identify one to three group of bars redundant, respectively. Design (a) eliminates all redundant bars identified in design (b) and (c) and removes $A_{37} - A_{39}$ summing up to nine redundant bars.

5.4   15-Bar Truss

This is a size and topology truss optimisation problem which is non-symmetric truss composed of 15 bars and truss undergoes three load cases and the sizing should be selected from a discrete set [36]. Figure 1 (III) shows the ground structure of this truss.

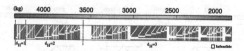

Fig. 4. Exact enumeration on 52-bar truss (right side truncated).

Table 3. Comparison of optimised designs for 52-bar truss.

	[35]	[22]	[25]	[18]	[16], [4], [7]	This Study		
						(a)	(b)	(c)
Best weight (kg)	1970.14	1906.76	1905.5	1904.830	1902.610	1862	1880.3	1869.7

Table 4 shows our findings by the proposed bilevel novelty search compared with other methods. We can see that designs (b) and (c) find the same weight, and they are symmetric around the vertical axis with respect to the topology and size of bars. Both designs remove 6 bars from the design space, and symmetrically they eliminate nodes 2 and 4 from the design space, respectively. Design (d) eliminates five bars and node 2 from the design space.

5.5   72-Bar Truss

72-bar truss represents a four storey structure for size and topology optimisation where which is a symmetric truss composed of 72 bars grouped into 16 groups of bars. The truss undergoes two load cases and the sizing of bars should be selected from a set [35]. Table 5 shows our findings by the proposed bilevel novelty search compared with other methods. We can see that designs (b) and (c) identify five groups of bars as redundant, including four common groups. This will lead to the elimination of 16 bars from the design space.

5.6   47-Bar Truss

This problem represents a transmission tower with symmetric truss bars grouped into 27 groups [13], and it considers size and shape optimisation in the lower level. Table 6 shows our findings in comparison with state of the art. We can see that our method obtained three different designs with respect to the topology. Design (a) incorporates 23 topology bars, and design (b) and (c) both incorporate 21 groups of bars and the main difference is including member groups of 23 and 26, respectively.

5.7   200-Bar Truss

This size and topology optimisation problem includes 200-bars grouped into 29 bar groups [18]. The truss undergoes three loading conditions and is subject to no displacement constraint. Table 7 shows our findings and compared with reported weights in state of the art. The best-reported weight is 25156.5 lb, but this is

Table 4. Comparison of optimised designs for 15-bar truss.

	[36]	[4]	This Study		
			(a)	(b), (c)	(d)
Best weight (kg)	142.12	105.74	89.899	90.223	91.874

Table 5. Comparison of optimised designs for 72-bar truss.

	[35]	[22]	[4], [7]	This Study		
				(a)	(b)	(c)
Best Weight (lb)	400.66	387.94	385.54	368.16	369.15	370.15

an infeasible design because it violates the stress constraint by about 8%. We can see that the best design obtained by our method (design (a)) outperforms other designs. Design (a) and (c) both include 24 group bars but with different topologies; however, Design (b) incorporates all group members as 29.

5.8   224-Bar Truss

This problem represents a pyramid truss where the truss bars are grouped into 32 groups and considers size, shape and topology optimisation subject to complex design specifications [13]. Table 8 lists obtained designs by our method compared with state of the art. Design (a) outperforms other designs, and design (b) is as alike as the optimum design obtained in [2].

5.9   68-Bar Truss

This size, shape and topology optimisation problem is a multi-load truss optimisation with 68 non-symmetric topology design variables [2]. The optimum design should be feasible considering the structural reactions subject to 8 different loading conditions. Table 9 shows our findings in comparison with state of the art. We can see that our method obtained the best design alike to the one obtained in [2], where this design incorporates 34 group bars. However, designs (b) and (c) include 37 and 39 bars, respectively.

Table 6. Comparison of optimised designs for 47-bar truss.

	[13]	[28]	[6]	[2]	This Study		
					(a)	(b)	(c)
Best weight (lb)	1885.070	1871.700	1836.462	1727.624	1724.947	1726.044	1727.624

Table 7. Comparison of optimised designs for 200-bar truss where † denotes the reported solution is infeasible.

	[4]	[16]	[18]	[7]	This Study		
					(a)	(b)	(c)
Best weight (lb)	27163.59	27858.50	25156.50†	27282.54	27144.0	27575.0	27744.0

Table 8. Comparison of optimised designs for 224-bar truss.

	[13]	[14]	[2]	This Study		
				(a)	(b)	(c)
Best weight (lb)	5547.500	4587.290	3079.446	3063.866	3079.446	3102.079

Table 9. Comparison of optimised designs for 68-bar truss.

	[28]	[2]	This Study		
			(a)	(b)	(c)
Best weight (lb)	1385.800	1166.062	1166.062	1167.528	1169.039

6   Conclusions

In this paper we considered bilevel optimisation of topology and size of trusses subject to discrete sizes. For the lower level optimisation, we use a reliable evolutionary optimiser. For the upper level, we employ exact enumeration and novelty-driven binary particle swarm optimisation to explore the upper level. We also use a repair mechanism to fix the infeasible solutions in the upper level that violate truss constraints.

In our experiments, we analysed the search space of smaller problems without randomness in the upper level. We also observed that we can find multiple distinct high-quality solutions with respect to the topology – moreover, we have found new best solutions for 8 out of 9 test problems.

Bilevel optimisation problems nest an optimisation problem into another where it increases the computational expense. This is the main drawback of this study. We also setup our algorithms with standard parameters. For future studies, it could be interesting (1) to investigate automated tuning of the algorithm, (2) to study this approach for large-scale trusses and (3) to improve it considering the computational expense of the problem.

References

1. Ahrari, A., Atai, A.A., Deb, K.: A customized bilevel optimization approach for solving large-scale truss design problems. Eng. Optim. **52**(12), 2062–2079 (2020)
2. Ahrari, A., Deb, K.: An improved fully stressed design evolution strategy for layout optimization of truss structures. Comput. Struct. **164**, 127–144 (2016)

3. Brooks, T.R., Kenway, G.K., Martins, J.R.: Undeflected common research model (UCRM): an aerostructural model for the study of high aspect ratio transport aircraft wings. In: 35th AIAA Applied Aerodynamics Conference, p. 4456 (2017)
4. Cheng, M.: A Hybrid Harmony Search algorithm for discrete sizing optimization of truss structure. Autom. Constr. **69**, 21–33 (2016)
5. Deb, K., Gulati, S.: Design of truss-structures for minimum weight using genetic algorithms. Finite Elem. Anal. Des. **37**(5), 447–465 (2001)
6. Degertekin, S.O., Lamberti, L., Ugur, I.: Sizing, layout and topology design optimization of truss structures using the JAVA algorithm. Appl. Soft Comput. **70**, 903–928 (2018)
7. Degertekin, S., Lamberti, L., Ugur, I.: Discrete sizing/layout/topology optimization of truss structures with an advanced JAVA algorithm. Appl. Soft Comput. **79**, 363–390 (2019)
8. Droste, S., Jansen, T., Wegener, I.: On the analysis of the (1+1) evolutionary algorithm. Theor. Comput. Sc. **276**(1), 51–81 (2002)
9. Fenton, M., McNally, C., Byrne, J., Hemberg, E., McDermott, J., O'Neill, M.: Automatic innovative truss design using grammatical evolution. Autom. Constr. **39**(C), 59–69 (2014)
10. Finotto, V.C., da Silva, W.R., Valášek, M., Štemberk, P.: Hybrid fuzzy-genetic system for optimising cabled-truss structures. Adv. Eng. Softw. **62**, 85–96 (2013)
11. Galvao, D.F., Lehman, J., Urbano, P.: Novelty-driven particle swarm optimization. In: Bonnevay, S., Legrand, P., Monmarché, N., Lutton, E., Schoenauer, M. (eds.) EA 2015. LNCS, vol. 9554, pp. 177–190. Springer, Cham (2016). https://doi.org/10.1007/978-3-319-31471-6_14
12. Hasancebi, O.: Optimization of truss bridges within a specified design domain using evolution strategies. Engi. Optim. **39**(6), 737–756 (2007)
13. Hasancebi, O., Erbatur, F.: Layout optimization of trusses using improved GA methodologies. Acta mechanica **146**(1), 87–107 (2001)
14. Hasançebi, O., Erbatur, F.: Layout optimisation of trusses using simulated annealing. Adv. Eng. Softw. **33**(7), 681–696 (2002)
15. He, S., Prempain, E., Wu, Q.: An improved particle swarm optimizer for mechanical design optimization problems. Eng. Optim. **36**(5), 585–605 (2004)
16. Ho-Huu, V., Nguyen-Thoi, T., Vo-Duy, T., Nguyen-Trang, T.: An adaptive elitist differential evolution for optimization of truss structures with discrete design variables. Comput. Struct. **165**(C), 59–75 (2016)
17. Islam, M.J., Li, X., Deb, K.: Multimodal truss structure design using bilevel and niching based evolutionary algorithms. In: Genetic and Evolutionary Computation Conference (GECCO). pp. 274–281. Association for Computing Machinery (2017)
18. Kaveh, A., Talatahari, S.: A particle swarm ant colony optimization for truss structures with discrete variables. J. Constr. Steel Res. **65**(8–9), 1558–1568 (2009)
19. Kennedy, J., Eberhart, R.: Particle swarm optimization. In: International Conference on Neural Networks (ICNN). vol. 4, pp. 1942–1948. IEEE (1995)
20. Khayyam, H., Jamali, A., Assimi, H., Jazar, R.N.: Genetic programming approaches in design and optimization of mechanical engineering applications. In: Jazar, R.N., Dai, L. (eds.) Nonlinear Approaches in Engineering Applications, pp. 367–402. Springer, Cham (2020). https://doi.org/10.1007/978-3-030-18963-1_9
21. Kicinger, R., Arciszewski, T., De Jong, K.: Evolutionary computation and structural design: a survey of the state-of-the-art. Comput. Struct. **83**(23–24), 1943–1978 (2005)
22. Lee, K.S., Geem, Z.W., Lee, S.h., Bae, K.w.: The harmony search heuristic algorithm for discrete structural optimization. Eng. Optim. **37**(7), 663–684 (2005)

23. Lehman, J., Stanley, K.O.: Exploiting open-endedness to solve problems through the search for novelty. In: ALIF, pp. 329–336. Citeseer (2008)
24. Lehman, J., Stanley, K.O.: Abandoning objectives: evolution through the search for novelty alone. Evol. Comput. **19**(2), 189–223 (2011)
25. Li, L.J., Huang, Z.B., Liu, F.: A heuristic particle swarm optimization method for truss structures with discrete variables. Comput. Struct. **87**(7–8), 435–443 (2009)
26. Li, X., Epitropakis, M.G., Deb, K., Engelbrecht, A.: Seeking multiple solutions: an updated survey on niching methods and their applications. IEEE Trans. Evol. Comput. **21**(4), 518–538 (2016)
27. Martinez, A.D., Osaba, E., Oregi, I., Fister, I., Fister, I., Ser, J.D.: hybridizing differential evolution and novelty search for multimodal optimization problems. In: Genetic and Evolutionary Computation Conference (GECCO) Companion, pp. 1980–1989 (2019)
28. Panagant, N., Bureerat, S.: Truss topology, shape and sizing optimization by fully stressed design based on hybrid grey wolf optimization and adaptive differential evolution. Eng. Optim. **50**(10), 1645–1661 (2018)
29. Rajeev, S., Krishnamoorthy, C.S.: Discrete optimization of structures using genetic algorithms. J. Struct. Eng. **118**(5), 1233–1250 (1992)
30. Rao, G.V.: Optimum designs for transmission line towers. Comput. Struct. **57**(1), 81–92 (1995)
31. Seber, G., Ran, H., Nam, T., Schetz, J., Mavris, D.: Multidisciplinary design optimization of a truss braced wing aircraft with upgraded aerodynamic analyses. In: 29th AIAA Applied Aerodynamics Conference, p. 3179 (2011)
32. Sinha, A., Malo, P., Deb, K.: A review on bilevel optimization: from classical to evolutionary approaches and applications. IEEE Trans. Evol. Comput. **22**(2), 276–295 (2017)
33. Stolpe, M.: Truss optimization with discrete design variables: a critical review. Struct. Multidiscipl. Optim. **53**(2), 349–374 (2015). https://doi.org/10.1007/s00158-015-1333-x
34. Topping, B.: Shape optimization of skeletal structures: a review. J. Struct. Eng. **109**(8), 1933–1951 (1983)
35. Wu, S.J., Chow, P.T.: Steady-state genetic algorithms for discrete optimization of trusses. Comput. Struct. **56**(6), 979–991 (1995)
36. Zhang, Y.N., Liu, P., Liu, B., Zhu, C.Y., Li, Y.: Application of improved hybrid genetic algorithm to optimized design of architecture structures. Huanan Ligong Daxue Xuebai(Ziran Kexue Ban)/J. South China Univ. Technol. (Natural Science Edition)(China) **33**(3), 69–72 (2005)

A Beam Search for the Shortest Common Supersequence Problem Guided by an Approximate Expected Length Calculation

Jonas Mayerhofer, Markus Kirchweger, Marc Huber[(⊠)], and Günther Raidl

TU Wien, Vienna, Austria
e01633065@student.tuwien.ac.at, {mk,mhuber,raidl}@ac.tuwien.ac.at

Abstract. The shortest common supersequence problem (SCSP) is a well-known NP-hard problem with many applications, in particular in data compression, computational molecular biology, and text editing. It aims at finding for a given set of input strings a shortest string such that every string from the set is a subsequence of the computed string. Due to its NP-hardness, many approaches have been proposed to tackle the SCSP heuristically. The currently best-performing one is based on beam search (BS). In this paper, we present a novel heuristic (AEL) for guiding a BS, which approximates the expected length of an SCSP of random strings, and embed the proposed heuristic into a multilevel probabilistic beam search (MPBS). To overcome the arising scalability issue of the guidance heuristic, a cut-off approach is presented that reduces large instances to smaller ones. The proposed approaches are tested on two established sets of benchmark instances. MPBS guided by AEL outperforms the so far leading method on average on a set of real instances. For many instances new best solutions could be obtained.

Keywords: Shortest Common Supersequence Problem · Beam Search

1 Introduction

We define a string s as a finite sequence of letters from a finite alphabet Σ. A subsequence of a string s is a sequence derived by deleting zero or more letters from that string without changing the order of the remaining letters. If s is a subsequence of another string x, then x is a supersequence of s. A common supersequence of a set of n non-empty strings $S = \{s_1, \ldots, s_n\}$ is a string x consisting of letters from Σ, which is a supersequence of each $s \in S$.

The *shortest common supersequence problem* (SCSP) asks for a shortest possible string that is a common supersequence of a given set of strings S. For example, a shortest common supersequence of the strings GAATG, AATGG, and TAATG

This project is partially funded by the Doctoral Program "Vienna Graduate School on Computational Optimization", Austrian Science Foundation (FWF), grant W1260-N35.

L. Pérez Cáceres and S. Verel (Eds.): EvoCOP 2022, LNCS 13222, pp. 127–142, 2022.
https://doi.org/10.1007/978-3-031-04148-8_9

is GTAATGG. The problem is not only of theoretical interest but has important applications in many areas, including data compression [21], query optimization in databases [20], AI planning [8], and bioinformatics [14,15]. Hence, there is a vast benefit when finding more effective algorithms for solving the SCSP effectively. For a fixed number of strings n, the SCSP can be solved in time $\mathcal{O}(m^n)$ by dynamic programming, where m is the length of the longest string [12]. As typical values for real instances go up to $n = 500$ and $m = 1000$ [10], this approach is not feasible in practice. For general n, the problem is known to be NP-hard [11], even under a binary alphabet [19] or strings with length two [23]. Although there exists an exact algorithm based on dynamic programming by Irving et al. [12], approximation and heuristic algorithms seem to be unavoidable tackling larger instances in practice. The currently best-performing one amongst them relies on beam search (BS) [10].

Beam search is a well-known incomplete tree search that explores a state graph by only expanding and further pursuing the most promising successor nodes of each level. Besides the SCSP, BS is also able to shine on similar problems like the Longest Common Subsequence Problem [6] and the Longest Common Palindromic Subsequence Problem [7]. For these problems, a BS guided by a theoretically derived function EX that approximates the expected length of the result of random strings from a partial solution achieved exceptional performance and outperformed other approaches on many instances.

Inspired by these approaches, we present a novel function that approximates the expected length of an SCSP of random strings and utilizes it as guidance heuristic in the multilevel probabilistic beam search (MPBS) from Gallardo et al. [10]. While our guidance heuristic performs well on smaller instances, numerical issues arise during its calculation on larger instances. To deal with this scalability issue, we present a cut-off approach that effectively reduces large instances to smaller ones. In our experimental evaluation, we consider a standard BS as well as the MPBS, both guided by the new expected value heuristic with and without the cut-off approach, and evaluate them on established benchmark instances. The experimental results show that our approaches can be highly effective and, in many cases, yield better solutions than the so far leading approach. On average, we are able to outperform the leading approach on a test set of large real instances and are competitive on a set of small random instances.

Section 2 reviews related work. In Section 3, we introduce some definitions and notations to describe the SCSP formally. The general BS framework for the SCSP with dominance check and the derivation of the new guidance heuristic as well as the cut-off approach are presented in Section 4. Experimental results of the methods on two established sets of benchmark instances, including some real-world instances, are compared to each other and the best-known results from the literature in Section 5. Finally, we conclude in Section 6, where we also outline promising future work.

2   Related Work

A first definition of the SCSP and an NP-hardness proof for an arbitrary number of sequences over an alphabet of size five were given by Maier in 1976 [16]. Later, in 1994, Irving et al. [12] solved the SCSP for a fixed number of strings in polynomial time by dynamic programming.

In addition to this exact algorithm, many approaches have been proposed to tackle the SCSP heuristically for large instances. A simple greedy heuristic called Majority Merge by Jiang et al. [13] constructs a supersequence by incrementally adding the symbol that occurs most frequently at the beginning of the strings in S and then deleting these symbols from the front of the corresponding strings. However, if the strings in S have different lengths, symbols at the beginning of shorter strings are as likely to be deleted. Therefore, Branke et al. [4,5] suggested a Weighted Majority Merge (WMM) that uses the string lengths as weights. A simple $|\Sigma|$-approximation algorithm called Alphabet was presented by Barone et al. [1]. For an alphabet $\Sigma = \{a_1, \ldots, a_\sigma\}$, this algorithm returns as trivial solution $(a_1 \cdot a_2 \cdot \ldots \cdot a_\sigma)^m$. A Deposition and Reduction algorithm was introduced by Ning et al. [18]. It first generates a small set of common supersequences and then tries to shorten these by deleting one or more symbols while preserving the common supersequence property. For generating a small set of common supersequences, a Look Ahead Sum Height (LA-SH) [14] and the Alphabet algorithm [1] are used. The LA-SH algorithm [14] extends the Majority Merge algorithm by a *look-ahead* strategy: Instead of considering only one step for choosing the best letter to add, the LA-SH algorithm looks several steps ahead before a letter is added.

In the context of BS, Gallardo et al. [9] presented a hybrid of a Memetic Algorithm with a BS, and Blum et al. [3] suggested a BS with a probabilistic approach for including elements in the beam, called Probabilistic Beam Search (PBS). More specifically, PBS calculates heuristic values using a look-ahead version of the WMM from [4,5] and computes the probability of a partial solution to be chosen based on these values. However, instead of always selecting a partial solution probabilistically, they employ a mixed strategy in which, at random, either the partial solution with the highest probability value is taken or a solution is chosen by a roulette-wheel selection. Furthermore, for reducing runtime, lower bounds are calculated by adding up the length of the partial solution x and the maximum number of occurrences of each symbol in any of the remaining strings. Later, Mousavi et al. [17] introduced a highly successful Improved BS (IBS) which outperformed all previous algorithms for solving the SCSP in all experiments they performed. This algorithm uses some basic laws of probability theory to calculate the probability that a uniform random string of a certain length is a supersequence of a set of strings S under the assumption that all strings in S are independent and some further simplifying assumptions. This probability is then used as heuristic value to guide the BS.

A multilevel PBS algorithm (MPBS) presented by Gallardo et al. [10] could achieve even better results than IBS and thus constitutes so far the state-of-the-art for the SCSP. The MPBS follows a destroy and repair paradigm.

First, an initial solution is generated by the PBS framework from Blum et al. [2] but using the guidance heuristic from the IBS. Afterwards, two processes called PBS-Perturbation and PBS-Reduction, are executed on the current solution until the allowed execution time is reached. PBS-Perturbation replaces a randomly chosen symbol of the current solution by a different one. This replacement can lead to an infeasible solution. Therefore, a repairing mechanism using PBS is employed, where PBS is run for the (now) uncovered part on the right side, and it is tried to find a shorter solution. PBS-Reduction first splits the current solution x into a prefix (x_L) and a suffix (x_R) and computes the longest suffix for each string in S that is a subsequence of x_R. The parts (prefixes) of all strings $s \in S$ that do not belong to the longest suffixes yield a new instance of the problem. Solving this instance using PBS yields a supersequence (x'_L) for the prefixes. If $|x'_L| < |x_L|$, then x_L is substituted by x'_L, i.e., $x = x'_L \cdot x_R$. This process is executed for all possible splits of the current solution and restarted with the enhanced solution as soon as an improvement is made.

3   Preliminaries

We denote the *length* of a string s by $|s|$, and the empty string by ε. If s is a string of length m, then $s[j]$ denotes the j-th letter of s for $1 \le j \le m$, and $s[j, j']$ refers to the substring $s[j]s[j+1] \ldots s[j']$ for $1 \le j \le j' \le m$ and the empty string else. Let s_1, s_2 be two strings, then $s_1 \cdot s_2$ denotes the concatenation of these. Furthermore, we denote by $s_1 \preceq s_2$ that s_1 is a subsequence of s_2 and s_2 a supersequence of s_1. Let $S = \{s_1, \ldots, s_n\}$ be a set of n strings and x another string. Then, by $p_i(x)$ we denote the largest integer such that $s_i \in S$, and $s_i[1, p_i(x)] \preceq x$. We call $p = (p_i)_{i=1,\ldots,n}$ *position vector* of a (partial) solution x as it indicates the already covered parts of the input strings. It further follows that x is a common supersequence of S if $|s_i| = p_i(x)$ for all $i \in \{1, \ldots, n\}$. Additionally, we define $r_i(x) = s_i[p_i(x) + 1, |s_i|]$ as the remaining part of the string s_i not covered by x. We call a string x a *partial solution* of the problem, if it is not (yet) a common supersequence.

4   Beam Search Framework

We now present the general BS framework with dominance check on which we build and utilize the basic ideas of IBS [17] and EX [6] to calculate the *approximate expected length* of an SCSP on random strings. Moreover, we describe a cut-off approach to make larger instances tractable without running into numerical issues.

Beam search is a prominent graph search algorithm that expands nodes in a breadth-first search manner to find a best path from a root node to a target node. To keep the computational effort within limits, BS evaluates the reached nodes at each level and selects a subset of only up to β most promising nodes to pursue further. The selected subset of nodes is called *beam B*, and parameter β *beam width*. In the context of the SCSP, the state graph is a directed acyclic graph $G = (V, A)$, where a state (node) $v \in V$ contains a position vector p^v, which represents the SCSP sub-instance with input strings $\{r_1^v, \ldots, r_n^v\}$ with $r_i^v = s_i[p_i^v + 1, |s_i|]$, $i = 1, \ldots, n$. The root node $r \in V$ has position vector $p^r = (1, \ldots, 1)$ and corresponds to the original SCSP instance. The terminal node $t \in V$ with $p^t = (|s_i|)_{i=1,\ldots,n}$ represents the instance with all strings being empty, i.e., all letters of the original instance have already been covered. An arc $(u, v) \in A$ refers to transitioning from state u to state v by appending a letter $a \in \Sigma$ to a partial solution, and thus, arc (u, v) is labeled by this letter denoted by $\ell(u, v) = a$. Appending a letter $a \in \Sigma$ is only considered feasible if this letter corresponds with the first letter of at least one remaining string r_i^u, $i = 1, \ldots, n$. Nodes are expanded until the terminal node $t \in V$ is reached. Then, the BS returns a shortest path from the root node to the terminal node, which represents the final valid solution. In general, there may exist several different paths from the root to some node, corresponding to different (partial) solutions yielding the same position vector. From these, we always only have to keep one shortest path. Thus, it is enough to actually store with each node a single reference to the parent node in order to finally derive the solution in a backward manner. For deciding which nodes are selected into the beam, a heuristic function is needed that expresses how promising each state is. We use for this purpose a new approximation of the expected length an optimal solution to the SCSP sub-instance induced by a state has, for short the Approximate Expected Length (AEL); it will be described in Section 4.3.

A pseudocode for the BS framework is shown in Algorithm 1. The algorithm starts with the root node. Function EXTENDANDEVALUATE expands each node of the current beam B by creating a new node for each feasible letter $a \in \Sigma$, and an arc, labeled a, between the original node and the new node, and it updates the p vector. Each new node is evaluated by calculating its AEL. In line 5, an optional dominance check is performed, which may discard nodes that are dominated by others; details follow in Sect. 4.1. Line 6 does the actual selection of the up to β best nodes according to the heuristic values to obtain the new beam.

Procedure EXTENDANDEVALUATE runs in time $\mathcal{O}(\beta |\Sigma| T_{\text{AEL}})$, where T_{AEL} is the time of one AEL calculation. The cardinality of the set of new nodes V_{ext} is at most $\beta |\Sigma|$. Sorting V_{ext} for selecting the beam takes $\mathcal{O}(\beta |\Sigma| \cdot \log(\beta |\Sigma|))$ time. The total runtime of the BS without dominance check is therefore

$$\mathcal{O}\left(l \beta |\Sigma| \cdot (T_{\text{AEL}} + \log(\beta |\Sigma|))\right), \tag{1}$$

where l is the length of the solution, i.e., levels of the BS. As we will argue, the dominance check we apply also does not increase this asymptotic time.

4.1 Dominance Check

A dominance check is used to filter out certain nodes that cannot be part of a shortest r–t path. We say a node u *dominates* node v at the current BS level if $p^u \neq p^v$ and $p_i^u \geq p_i^v$ for all $i = 1, \ldots, n$. Nodes that are dominated

---

Algorithm 1. BS for the SCSP

   Input: instance (S, Σ)
   Output: a common supersequence of S
1: $B \leftarrow \{p^r\}$ ▷ beam
2: **while** true **do**
3:    $V_{\text{ext}} \leftarrow$ EXTENDANDEVALUATEB
4:    **if** t $\in V_{\text{ext}}$ **then return** solution corresponding to r–t path
5:    $V_{\text{ext}} \leftarrow$ DOMINANCECHECKV_{ext} ▷ optional
6:    $B \leftarrow$ select (up to) β best nodes from V_{ext}

---

by other nodes at the current level can be discarded as they cannot lead to better solutions. More specifically, we apply the restricted κ-dominance check in the spirit of [17] to avoid a quadratic effort in the number of nodes. Let $K \subseteq V_{\text{ext}}$ be the subset of the (up to) κ best-ranked nodes according to the heuristic evaluation; $\kappa \geq 0$ is hereby a strategy parameter. We then do pairwise domination checks only among V_{ext} and K: First, the expanded set of nodes V_{ext} is sorted in non-decreasing order by the nodes' heuristic values, and the leftmost κ solutions $K \subseteq V_{\text{ext}}$ are selected. If $v \in V_{\text{ext}}$ is dominated by any $u \in K$, then node v is discarded. Note that in contrast to [17], dominance within the leftmost κ solutions is also checked according to their order. A single dominance check of two nodes takes time $\mathcal{O}(n)$, and therefore, the whole κ-dominance check for one level is done in time $\mathcal{O}(\kappa \beta |\Sigma| n)$. If we consider κ to be a constant, the BS's total asymptotic time complexity (1) does not change.

4.2 Heuristic Function from IBS

Before we introduce AEL, we review the heuristic function from Mousavi et al. [17] as it provides a basis for our considerations. Its basic idea is to calculate the probability of a random string of length k being a common supersequence of an independent random string of length q.

Theorem 1. *Let w, y be two uniform random strings with length q and k respectively. The probability of y being a common supersequence of w is*

$$\mathbb{P}(q, k) = \begin{cases} 1 & \text{if } q = 0 \\ 0 & \text{if } q > k \\ \frac{1}{|\Sigma|}\mathbb{P}(q-1, k-1) + \frac{|\Sigma|-1}{|\Sigma|}\mathbb{P}(q, k-1) & \text{otherwise.} \end{cases} \quad (2)$$

By using Theorem 1, the probability that a random string y is a common supersequence of a set of strings S can be calculated by $\prod_{s \in S} \mathbb{P}(|s|, |y|)$. Hence, for a partial solution x, the probability that y is a common supersequence of all $r_i(x)$ is $h(x) = \prod_{i=1}^{n} \mathbb{P}(|r_i(x)|, |y|)$. Mousavi et al. directly use this probability as guidance function with the length of the string y calculated by

$$|y| = \max_{v \in V_{\text{ext}}, \, i \in \{1, \dots, n\}} |r_i^v| \cdot \log(|\Sigma|), \quad (3)$$

but also mention that selecting the "best" length would need further investigation.

4.3   Approximate Expected Length (AEL)

Now, we present our new guidance heuristic AEL, inspired by the earlier work for the longest common subsequence problem from Djukanovic et al. [6] that approximates the expected length of an SCS on uniform random strings. Let Y be the random variable corresponding to the length of an SCSP of n uniformly randomly generated strings S, and let $Y_k \in \{0, 1\}$ be a binary random variable indicating if there is a supersequence of length k for S. The realizations of Y cannot be larger than $u = |\Sigma| m$ and smaller than m, where $m = \max_{s \in S} |s|$. The upper bound can be trivially shown by taking an arbitrary permutation of the alphabet and repeating it m times, cf. [22]. These definitions enable us to express $\mathbb{E}(Y)$ in terms of $\mathbb{E}(Y_k)$ by using some basic laws from probability theory:

$$
\begin{aligned}
\mathbb{E}(Y) &= \sum_{k=m}^{u} k \, \mathbb{P}(Y = k) \\
&= \sum_{k=m}^{u} k \cdot (\mathbb{E}(Y_k) - \mathbb{E}(Y_{k-1})) \\
&= u \, \mathbb{E}(Y_u) - \sum_{k=m}^{u-1} \mathbb{E}(Y_k) - m \, \mathbb{E}(Y_{m-1}) \\
&= u - \sum_{k=m}^{u-1} \mathbb{E}(Y_k).
\end{aligned}
\tag{4}
$$

Assume that the probability of a sequence being a common supersequence of all strings in S is independent for distinct sequences. The probability that a string of length k is a common supersequence of S is then given by $\prod_{i=1}^{n} \mathbb{P}(|s_i|, k)$, and the probability that this is not the case by $1 - \prod_{i=1}^{n} \mathbb{P}(|s_i|, k))$. Under the assumption that these probabilities are independent for all $|\Sigma|^k$ possible strings of length k, the probability that none of them is a common supersequence is $(1 - \prod_{i=1}^{n} \mathbb{P}(|s_i|, k))^{|\Sigma|^k}$. Hence, $\mathbb{E}(Y_k) = 1 - (1 - \prod_{i=1}^{n} \mathbb{P}(|s_i|, k))^{|\Sigma|^k}$ holds under these simplifying assumptions. This enables us to utilize the probability function $\mathbb{P}(q, k)$ from the previous Sect. 4.2, yielding

$$
\mathbb{E}(Y) = u - \sum_{k=m}^{u-1} \left(1 - (1 - \prod_{i=1}^{n} \mathbb{P}(|s_i|, k))^{|\Sigma|^k} \right).
\tag{5}
$$

To avoid numerical problems and speed up the computation, the expression is not evaluated directly. In practice, most of the terms $1 - (1 - \prod_{i=1}^{n} \mathbb{P}(|s_i|, k))^{|\Sigma|^k}$ are either quite close to zero or one. Moreover, it is easy to see that $1 - (1 - \prod_{i=1}^{n} \mathbb{P}(|s_i|, k_1))^{|\Sigma|^{k_1}} \leq 1 - (1 - \prod_{i=1}^{n} \mathbb{P}(|s_i|, k_2))^{|\Sigma|^{k_2}}$ holds for $k_1 < k_2$. Hence, a divide-and-conquer approach is employed to calculate the sum in Eq. 5. More specifically, all values greater than $1 - \delta$ or smaller than δ are approximated with 0 or 1. We chose $\delta = 10^{-20}$, since in our experiments this value is small enough to have no major impact on the calculation of the expected length.

Since $|\Sigma|^k$ cannot be represented in a standard floating-point arithmetic for large k, the expression $(1 - \prod_{i=1}^{n} \mathbb{P}(|s_i| \preceq k))^{|\Sigma|^k}$ is decomposed into

$$\underbrace{\left(\left(\cdots (1 - \prod_{i=1}^{n} \mathbb{P}(|s_i|, k))^{|\Sigma|^p} \cdots \right)^{|\Sigma|^p} \right)^{|\Sigma|^{(k \bmod p)}}}_{\lfloor k/p \rfloor \text{ times}} \tag{6}$$

for $p = 10$. Additionally, if the product over the probabilities is very small, numerical issues occur. To tackle this problem, the expression in Equation 6 is estimated by using a Taylor series approximation if the product is smaller than a certain threshold. See [6] for more details on this.

Cut-Off. For larger instances, the heuristic values the above calculation yields are nevertheless often integers due to numerical imprecisions, because either $|\Sigma|^k$ is far too large or the product is far too small. This often results in many equal heuristic values for the nodes, and thus a poor differentiation of how promising the nodes are. To deal with this issue, we shorten the strings for the calculation when they exceed a certain length. More specifically, at each iteration of the while-loop in Algorithm 1, the length m_{ext} of the longest remaining string over all nodes in V_{ext} is taken, i.e., $m_{\text{ext}} = \max_{v \in V_{\text{ext}}, i=1,\ldots,n} |r_i^v|$, and a *cut-off* $C = \max(0, m_{\text{ext}} - \gamma)$ is determined, where γ is a strategy parameter. Instead of calculating AEL for a node $v \in V_{\text{ext}}$ over the original remaining string lengths $|r_i^v|$, it is now determined over the lengths $\max(0, |r_i^v| - C)$, $i = 1, \ldots, n$.

Computational Complexity. In the worst case, for each k, the n probabilities for all n input strings have to be multiplied and the stable power applied. Thus, AEL can be performed in time $\mathcal{O}((q|\Sigma|)^2 n/p)$, where $q = \max_{i=1,\ldots,n}(|r_i^v|)$. In practice, due to the divide-and-conquer approach and approximation of values close to zero and one, only a small fraction of the stable powers and multiplications is needed.

5   Experimental Evaluation

We implemented the BS as well as MPBS, both with AEL as guidance function, in Julia 1.6.2. All tests were performed on an Intel Xeon E5-2640 processor with 2.40 GHz in single-threaded mode and a memory limit of 8GB. We compare these two approaches among each other and to results from IBS and the original MPBS as reported in [10].

5.1   Test Instances

Two benchmark sets already used in [10] are considered to evaluate the approaches. The first set denoted as Real consists of real-world instances, and the second one, **Rand**, of random instances. Set **Rand** consists of five instances for each $|\Sigma| \in \{2, 4, 8, 16, 24\}$, each instance having four random strings of length 40 plus four random strings of length 80, i.e., $n = 8$. Set **Real** consists of real DNA and protein instances. There are ten DNA instances for each combination of $n \in \{100, 500\}$ equally long strings of length $m \in \{100, 500, 1000\}$, and $|\Sigma| = 4$, in total thus 60 instances. Moreover, there are 10 protein instances for each $(n, m) \in \{(100, 100), (500, 100), (100, 500)\}$, where again all strings are equally long, and $|\Sigma|$ varies in $\{20, \ldots, 24\}$; in total these are 30 instances.

For both instance sets **Real** and **Rand**, a collection of so far best-known solution lengths l_{best} is provided in [10]. All instances are available online[1].

5.2   Impact of Cut-Off and Comparison to IBS

In this Section, we investigate the impact of the cut-off parameter γ and compare results to the IBS from [17]. Unfortunately, we could not reproduce the results stated by Mousavi et al. in [17] due to some missing details in their paper. An inquiry was without success. Also, we could not obtain the original source codes of former approaches. Moreover, note that the results reported for IBS in the newer publication [10] differ from those in [17]. For us, this does not matter much since for the cases where our approaches performed better they do so for the results reported in both papers. We use the newer values reported in [10] for the comparison here. To enable comparability of our results the same beam width of $\beta = 100$ and the dominance check with $\kappa = 7$ was applied as in [10].

Preliminary tests for the cut-off parameter γ indicated that results are not particularly sensitive as long as $20 \leq \gamma \leq 40$. Higher values lead frequently to the numerical issues we want to avoid, while lower values often result in an over-simplification and poor guidance. We were not able to find a clear relationship of good values for γ, the number of input strings, and the alphabet size. We therefore investigate these two border values as well as calculating AEL without cut-off. The BS variants with these guidance heuristics are denoted in the following by BS_{AEL}, $\text{BS}_{\text{AEL},\gamma=20}$, $\text{BS}_{\text{AEL},\gamma=40}$, respectively.

Tables 1 and 2 summarize obtained results for **Real** and **Rand** and those reported in [10] for IBS. Listed are for each instance group average solution lengths $\bar{l}$, the respective standard deviation σ and the average runtime $\bar{t}$ in seconds. In each row, the best result w.r.t. solution length is printed bold.

The results show that BS_{AEL} performs worse than IBS in most cases. This is primarily caused by the many ties the numerical calculating of AEL yields. By using the cut-off approach with $\gamma \in \{20, 40\}$, the performance of our BS with AEL increases significantly, and we obtain better results than IBS on almost all instances on **Real**. The approaches $\text{BS}_{\text{AEL},\gamma \in \{20, 40\}}$ each outperform IBS on seven out of nine **Real** instance classes on average, where $\text{BS}_{\text{AEL},\gamma=\{40\}}$ outperforms IBS on all DNA instance classes, and $\text{BS}_{\text{AEL},\gamma=\{20\}}$ on two out of three protein instance classes. On the **Rand** instances, BS_{AEL} is almost on par with

---

[1] https://www.ac.tuwien.ac.at/research/problem-instances/.

Table 1. Real instances: average solution lengths and runtimes obtained by BS_{AEL}, $BS_{AEL,\gamma\in\{20,40\}}$, and results of IBS from [10]; $\beta = 100$.

	n	m	BS_{AEL}			$BS_{AEL,\gamma=20}$			$BS_{AEL,\gamma=40}$			IBS
			$\bar{l}$	σ_l	$\bar{t}[s]$	$\bar{l}$	σ_l	$\bar{t}[s]$	$\bar{l}$	σ_l	$\bar{t}[s]$	$\bar{l}$
DNA	100	100	272.4	2.7	2.2	**270.3**	2.2	2.0	271.6	2.6	2.3	272.3
DNA	500	100	287.9	1.8	6.2	**285.7**	1.5	5.1	287.5	2.4	6.1	288.1
DNA	100	500	1290.9	9.7	6.9	1279.7	5.5	5.3	**1279.2**	5.7	7.2	1284.6
DNA	500	500	1362.4	8.9	26.0	1342.9	7.4	28.7	**1341.2**	7.2	27.1	1351.6
DNA	100	1000	2567.9	26.6	8.6	2545.4	10.9	13.2	**2536.5**	11.8	11.8	2540.1
DNA	500	1000	2682.8	25.2	48.9	2654.2	21.5	46.5	**2641.6**	18.5	52.5	2662.9
PROTEIN	100	100	920.9	16.0	16.7	**896.0**	12.5	21.5	912.9	8.9	16.5	910.6
PROTEIN	500	100	1122.5	26.6	99.9	**1071.4**	16.2	87.2	1092.3	10.7	102.2	1118.1
PROTEIN	100	500	4473.8	86.8	84.8	4405.7	47.5	98.6	4434.0	32.4	85.1	**4374.9**

Table 2. Rand instances: average solution lengths and runtimes obtained by BS_{AEL}, $BS_{AEL,\gamma\in\{20,40\}}$ and results of IBS from [10]; $\beta = 100$.

$|\Sigma|$	BS_{AEL}			$BS_{AEL,\gamma=20}$			$BS_{AEL,\gamma=40}$			IBS
	$\bar{l}$	σ_l	$\bar{t}[s]$	$\bar{l}$	σ_l	$\bar{t}[s]$	$\bar{l}$	σ_l	$\bar{t}[s]$	$\bar{l}$
2	**109.4**	1.7	1.4	109.6	2.0	1.3	109.8	1.9	0.8	**109.4**
4	143.0	1.4	1.2	144.0	1.3	1.7	143.2	2.2	1.5	**142.4**
8	**180.4**	2.5	1.9	186.6	1.0	1.8	182.4	2.6	1.8	180.6
16	238.6	4.7	2.0	241.4	6.0	1.9	237.2	4.2	2.0	**235.6**
24	274.4	2.7	1.1	279.0	4.8	2.5	273.8	5.3	1.3	**268.8**

IBS, but, the cut-off extension did worsen instead of improving the average solution quality. The reason seems to be the strong differences in the lengths of the input strings – remember that in each instance, half of the strings only has half the length of the others. Moreover, string lengths are generally smaller than in the Real instances, and therefore fewer ties occur in BS_{AEL} without cut-off. We remark that the runtimes reported in [10] for IBS are in the same order of magnitude, but generally smaller than the ones observed for our approaches. However, these times can hardly be compared due to the different programming languages and hardware used.

5.3 Integrating AEL into MPBS

As MPBS [10] is the approach yielding so far the best results on average, we now equip it with AEL as guidance heuristic. We set the strategy parameters as in [10]: $\beta_{init} = 100$ for generating the initial solutions, $\beta_{pert} = 700$ for the perturbation, and $\beta_{redu} = 200$ for the reduction. The number of generated initial solutions was set to $\zeta_{init} = 3$, the number of perturbations per iteration to $\zeta_{init} = 7$, and $\lambda = 0.6$ giving the degree of randomness of PBS (i.e. 60% of all nodes are selected randomly from V_{ext} instead of taking the best node); these

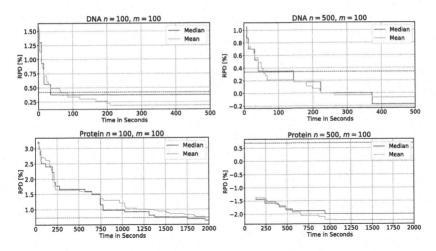

Fig. 1. RPD of $MPBS_{AEL,\gamma=20}$ over time in seconds. The dashed line marks the average RPD value from $MPBS_{IBS}$ [10].

values are not provided in [10] and therefore chosen by us following preliminary tests to balance exploration and exploitation.

In conjunction with AEL as guidance heuristic, we found that in the PBS-Reduction approach it seems to be better to not always increase the length of the prefix x_L of the current solution by one. Instead, we increase it by 5% of the current solution length. In this way, we had never more than 21 reduction runs per iteration, including a reduction run with the total length. This should not have a significant impact on the solution quality as long as the number of letters to be added is sufficiently small. The allowed runtime for MPBS was set to 500 s for the DNA instances and to 2000 s for the protein instances and 300 s for the Rand instances.

We denote MPBS with our AEL-heuristic without cut-off by $MPBS_{AEL}$, with cut-off for $\gamma \in \{20, 40\}$ by $MPBS_{AEL,\gamma=20}$ and $MPBS_{AEL,\gamma=40}$, respectively, and with the heuristic function used in IBS by $MPBS_{IBS}$. Figure 1 shows for $MPBS_{AEL,\gamma=20}$ how the mean and median solution length over all 10 instances of an instance group change over time. Instead of providing the average solution lengths, we present the relative percentage difference (RPD) which is obtained by $100\% \cdot (l - l_{best})/l_{best}$, where l denotes the solution length and l_{best} the best-known length as listed in [10]. It turned out that the reductions and perturbations decrease the solution length frequently at the beginning, but after some time improvements could be achieved only rarely.

Figure 2 shows the RPD for DNA and protein instances obtained by BS_{AEL}, $BS_{AEL,\gamma\in\{20,40\}}$, $MPBS_{AEL}$, and $MPBS_{AEL,\gamma\in\{20,40\}}$ as boxplots. We can observe here once more that $BS_{AEL,\gamma\in\{20,40\}}$ perform significantly better than the BS_{AEL} without cut-off. Further, turning to MPBS with its reductions and perturbations further improves the results in almost all cases significantly. For some instances, the results are already better than the previously best-known ones.

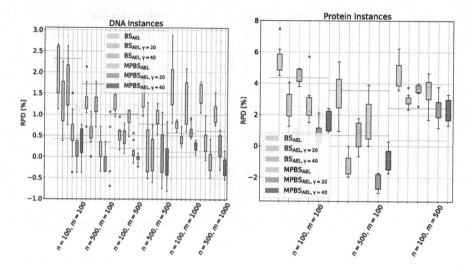

Fig. 2. Boxplots of the RPD for AEL and MPBS$_{AEL}$ on DNA and Protein instances. Dotted lines mark the average RPD from MPBS$_{IBS}$ [10] and dashed lines the average RPD from IBS [10].

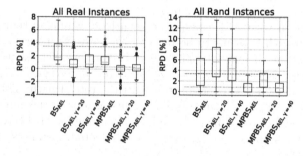

Fig. 3. Boxplots of the RPD for AEL and MPBS$_{AEL}$ over all **Real** and all **Rand** instances. Dotted lines mark the average RPD value from MPBS$_{IBS}$ [10] and dashed lines the average RPD from IBS [10].

Figure 3 shows the RPD over all instances of **Real** and **Rand**. For **Real**, MPBS$_{AEL,\gamma\in\{20,40\}}$ outperforms IBS and is close to MPBS$_{IBS}$. Furthermore, the figure shows that the average solution length of MPBS$_{AEL,\gamma\in\{20,40\}}$ lies below the one of MPBS$_{IBS}$ and thus MPBS$_{AEL,\gamma\in\{20,40\}}$ outperforms MPBS$_{IBS}$. We remark that a comparison on an instance-based basis is not possible as no values per instance are reported for MPBS$_{IBS}$ in [10]. Concerning **Rand**, we observe that BS$_{AEL}$ has almost the same average solution quality as IBS. Also MPBS$_{AEL}$, MPBS$_{AEL,\gamma=40}$ and MPBS$_{IBS}$ have a similar solution quality.

Tables 3 and 4 list average solution lengths and standard deviations of the MPBS variants including the results from [10]. Similarly to Table 1 and 2, we observe that for small random instances our approach does not improve on the previous approach while for almost all larger instances, we could achieve better

Table 3. Average solution lengths and standard deviations of the MPBS variants obtained on the Rand instances; results for $MPBS_{IBS}$ from [10]; runtime: 300 s.

$\lvert\Sigma\rvert$	$MPBS_{AEL}$		$MPBS_{AEL,\gamma=20}$		$MPBS_{AEL,\gamma=40}$		$MPBS_{IBS}$
	$\bar{l}$	σ_l	$\bar{l}$	σ_l	$\bar{l}$	σ_l	$\bar{l}$
2	109.0	1.7	109.4	1.6	109.0	1.7	**108.8**
4	139.8	1.6	140.4	1.9	**139.6**	1.5	139.8
8	177.6	1.5	180.6	1.4	177.8	1.9	**177.2**
16	**227.6**	3.9	232.2	3.9	228.0	1.7	227.7
24	257.6	3.0	264.2	2.8	259.6	3.3	**257.3**

Table 4. Average solution lengths and standard deviations of the MPBS variants obtained on the Real instances; results for $MPBS_{IBS}$ from [10]; runtime: 500 s for DNA instances, 2000 s for protein instances.

	n	m	$MPBS_{AEL}$		$MPBS_{AEL,\gamma=20}$		$MPBS_{AEL,\gamma=40}$		$MPBS_{IBS}$
			$\bar{l}$	σ_l	$\bar{l}$	σ_l	$\bar{l}$	σ_l	$\bar{l}$
DNA	100	100	268.6	1.7	**267.5**	1.6	267.9	1.8	268.1
DNA	500	100	284.6	1.7	**283.7**	1.6	284.8	1.1	285.0
DNA	100	500	1286.1	10.0	1275.1	6.1	**1272.8**	6.2	1276.0
DNA	500	500	1353.1	6.4	1341.1	8.0	**1338.6**	6.6	1343.1
DNA	100	1000	2559.2	25.4	2541.5	10.6	2532.3	11.0	**2529.1**
DNA	500	1000	2672.5	23.1	2649.9	21.1	**2637.9**	20.3	2647.9
PROTEIN	100	100	897.5	17.6	**880.0**	13.0	887.7	8.2	880.2
PROTEIN	500	100	1103.7	26.2	**1060.5**	14.7	1075.9	11.9	1092.2
PROTEIN	100	500	4430.9	70.3	4382.6	51.8	4378.1	41.1	**4296.7**

results. The approaches $MPBS_{AEL,\gamma\in\{20,40\}}$ each outperform $MPBS_{IBS}$ on six out of nine Real instance classes on average. In total only one DNA and one protein instance class remain where $MPBS_{IBS}$ was better on average.

5.4 Improving the Best-Known Solutions

By using a significantly larger beam width than 100, i.e., beam width $\beta \in \{1000, 2000, 5000, 10000\}$ and a timeout of eight hours per instance, we were able to improve several so far best-known solutions. In these experiments we considered BS_{AEL}, and $BS_{AEL,\gamma\in\{20,40,50\}}$. Also, we consider $MPBS_{AEL}$, and $MPBS_{AEL,\gamma\in\{20,40,50\}}$, with the settings given in Sect. 5.3, but with a timeout of eight hours per instance. Table 5 list for each instance group from Real the approach that achieved the best average solution length together with this solution value. In the case of ties, we list the fastest approach for BS_{AEL}, i.e., the one with smaller beam width, or all approaches in case of MPBS. For comparison, the table also lists the so far best-known average solution lengths from [10].

Table 5. Average solution lengths on **Real** with BS_{AEL}, $BS_{AEL,\gamma \in \{20,40,50\}}$, $MPBS_{AEL}$, $MPBS_{AEL,\gamma \in \{20,40,50\}}$ and high beam widths given in the table in comparison to the previously best-known results.

	n	m	method	β	best-found		best-known	
					$\bar{l}$	σ_l	$\bar{l}_{best}$	σ_l
DNA	100	100	$MPBS_{AEL,\gamma=20}$	-	**266.0**	1.8	267.0	1.8
DNA	500	100	$MPBS_{AEL,\gamma=20}$	-	**282.3**	0.9	284.0	1.9
DNA	100	500	$BS_{AEL,\gamma=50}$	10000	**1269.4**	6.4	1273.5	7.2
DNA	500	500	$MPBS_{AEL,\gamma \in \{40,50\}}$	-	**1333.1**	6.3	1340.6	4.4
DNA	100	1000	$BS_{AEL,\gamma=50}$	10000	**2521.4**	11.4	2526.3	12.5
DNA	500	1000	$BS_{AEL,\gamma=50}$	10000	**2631.1**	18.3	2644.5	22.3
PROTEIN	100	100	$BS_{AEL,\gamma=20}$	5000	**850.6**	9.3	873.8	8.5
PROTEIN	500	100	$BS_{AEL,\gamma=20}$	5000	**1040.7**	14.9	1084.7	14.4
PROTEIN	100	500	$BS_{AEL,\gamma=40}$	5000	**4201.2**	38.2	4280.0	44.5

Table 6. Average solution lengths on **Rand** with BS_{AEL} and high beam widths given in the table in comparison to the previously best-known results.

$|\Sigma|$	β	best-found		best-known	
		$\bar{l}$	σ_l	$\bar{l}_{best}$	σ_l
2	1000	**108.8**	1.3	**108.8**	1.3
4	2000	**139.2**	1.2	**139.2**	1.2
8	5000	**176.0**	1.6	**176.0**	1.7
16	5000	**222.4**	3.3	224.6	3.4
24	5000	**249.4**	1.9	252.4	1.6

Table 6 list the results for **Rand** with BS_{AEL}, since no other method yields better results. Our new solutions can be found online (see Footnote 1).

For all the instance groups, better or the same average results could be found. The table also shows that a higher beam width yields better results. Further, we can see that even for small γ the algorithm performs relatively well and sometimes even returns better results than for higher γ values, especially for the protein instances. This might be due to the larger alphabet size in comparison to the DNA instances. In total, we found new on average best solutions for all instance classes of **Real**, three times with $MPBS_{AEL}$ with different γ values, three times with $BS_{AEL,\gamma=50}$, two times with $BS_{AEL,\gamma=20}$ and once with $BS_{AEL,\gamma=40}$. For the **Rand** instance classes, we found the same average solutions for three out of five instance classes and new on average best solutions for the other two instance classes with a higher alphabet size of $|\Sigma| \in \{16, 24\}$.

6 Conclusions and Future Work

Inspired by previous work for the longest common subsequence problem, we developed a novel approximate expected length calculation for the SCSP and used it to guide a respective BS heuristic as well as the more advanced MPBS from [10]. To reduce ties that arise from numerical imprecisions in case of larger instances and impact performance, an effective cut-off approach was introduced. Our experiments show that by applying AEL instead of the heuristic used in IBS [17], we were able to achieve better results on most of the benchmark instances and to outperform MPBS$_{IBS}$, the so-far leading approach for the SCSP, on average. While AEL works particularly well on benchmark set `Real`, `Rand` instances turned out to be more challenging for AEL due to their particularity that the input strings differ heavily in their lengths. By using AEL in conjunction with higher beam widths and allowing longer runtimes, we could ultimately find new best-known solutions for almost all benchmark instances.

In future work, it would be interesting to adapt AEL for different variants of the SCSP, such as the constrained SCSP, or to apply machine learning techniques, in particular, reinforcement learning based, in order to learn a possibly even more appropriate guidance heuristic.

References

1. Barone, P., Bonizzoni, P., Vedova, G.D., Mauri, G.: An approximation algorithm for the shortest common supersequence problem: an experimental analysis. In: Proceedings of the 2001 ACM Symposium on Applied Computing, p. 56–60. ACM (2001)
2. Blum, C., Blesa, M.J.: Probabilistic beam search for the longest common subsequence problem. In: Stützle, T., Birattari, M., H. Hoos, H. (eds.) SLS 2007. LNCS, vol. 4638, pp. 150–161. Springer, Heidelberg (2007). https://doi.org/10.1007/978-3-540-74446-7_11
3. Blum, C., Cotta, C., Fernández, A.J., Gallardo, J.E.: A probabilistic beam search approach to the shortest common supersequence problem. In: Cotta, C., van Hemert, J. (eds.) EvoCOP 2007. LNCS, vol. 4446, pp. 36–47. Springer, Heidelberg (2007). https://doi.org/10.1007/978-3-540-71615-0_4
4. Branke, J., Middendorf, M.: Searching for shortest common supersequences by means of a heuristic-based genetic algorithm. In: Proceedings of the Second Nordic Workshop on Genetic Algorithms, pp. 105–113. University of Vaasa, Finland (1996)
5. Cotta, C.: A comparison of evolutionary approaches to the shortest common supersequence problem. In: Cabestany, J., Prieto, A., Sandoval, F. (eds.) IWANN 2005. LNCS, vol. 3512, pp. 50–58. Springer, Heidelberg (2005). https://doi.org/10.1007/11494669_7
6. Djukanovic, M., Raidl, G.R., Blum, C.: A beam search for the longest common subsequence problem guided by a novel approximate expected length calculation. In: Nicosia, G., Pardalos, P., Umeton, R., Giuffrida, G., Sciacca, V. (eds.) LOD 2019. LNCS, vol. 11943, pp. 154–167. Springer, Cham (2019). https://doi.org/10.1007/978-3-030-37599-7_14
7. Djukanovic, M., Raidl, G.R., Blum, C.: Anytime algorithms for the longest common palindromic subsequence problem. Comput. Oper. Res. **114**, 104827 (2020)

8. Foulser, D.E., Li, M., Yang, Q.: Theory and algorithms for plan merging. Artif. Intell. **57**(2–3), 143–181 (1992)
9. Gallardo, J.E., Cotta, C., Fernandez, A.J.: On the hybridization of memetic algorithms with branch-and-bound techniques. IEEE Trans. Syst. Man Cyber. Part B **37**(1), 77–83 (2007)
10. Gallardo, J.E.: A multilevel probabilistic beam search algorithm for the shortest common supersequence problem. PLoS ONE **7**(12), e52427 (2012)
11. Garey, M.R., Johnson, D.S.: Computers and Intractability; A Guide to the Theory of NP-Completeness. W. H. Freeman & Co., USA (1990)
12. Irving, R.W., Fraser, C.B.: Maximal common subsequences and minimal common supersequences. In: Crochemore, M., Gusfield, D. (eds.) CPM 1994. LNCS, vol. 807, pp. 173–183. Springer, Heidelberg (1994). https://doi.org/10.1007/3-540-58094-8_16
13. Jiang, T., Li, M.: On the approximation of shortest common supersequences and longest common subsequences. SIAM J. Comput. **24**(5), 1122–1139 (1995)
14. Kang, N., Pui, C.K., Wai, L.H., Louxin, Z.: A post-processing method for optimizing synthesis strategy for oligonucleotide microarrays. Nucleic Acids Res. **33**(17), e144–e144 (2005)
15. Kasif, S., Weng, Z., Derti, A., Beigel, R., DeLisi, C.: A computational framework for optimal masking in the synthesis of oligonucleotide microarrays. Nucleic Acids Res. **30**(20), e106 (2002)
16. Maier, D.: The complexity of some problems on subsequences and supersequences. J. ACM **25**(2), 322–336 (1978)
17. Mousavi, S.R., Bahri, F., Tabataba, F.: An enhanced beam search algorithm for the shortest common supersequence problem. Eng. Appl. Artif. Intell. **25**(3), 457–467 (2012)
18. Ning, K., Leong, H.W.: Towards a better solution to the shortest common supersequence problem: the deposition and reduction algorithm. BMC Bioinf. **7**(S4), S12 (2006)
19. Räihä, K., Ukkonen, E.: The shortest common supersequence problem over binary alphabet is NP-complete. Theor. Comput. Sci. **16**, 187–198 (1981)
20. Sellis, T.K.: Multiple-query optimization. ACM Trans. Database Syst. **13**(1), 23–52(1988)
21. Storer, J.A.: Data Compression: Methods and Theory. Press Computer Science, London (1988)
22. Sven, R.: The shortest common supersequence problem in a microarray production setting. Bioinformatics **19**, ii156–ii161 (2003)
23. Timkovskii, V.G.: Complexity of common subsequence and supersequence problems and related problems. Cybernetics **25**(5), 565–580 (1990)

Modeling the Costas Array Problem in QUBO for Quantum Annealing

Philippe Codognet[(✉)]

JFLI, CNRS/Sorbonne University/University of Tokyo,
7-3-1 Hongo, Bunkyo-ku, Tokyo 113-8656, Japan
codognet@is.s.u-tokyo.ac.jp

Abstract. We present experiments in solving constrained combinatorial optimization problems by means of Quantum Annealing. We describe how to model a hard combinatorial problem, the Costas Array Problem, in terms of QUBO (Quadratic Unconstrained Binary Optimization). QUBO is the input language of quantum computers based on quantum annealing such as the D-Wave systems and of the "quantum-inspired" special-purpose hardware such as Fujitsu's Digital Annealing Unit or Hitachi's CMOS Annealing Machine. We implemented the QUBO model for the Costas Array Problem on several hardware solvers based on quantum annealing (D-Wave Advantage, Fujitsu DA3 and Fixstars AE) and present some performance result for these implementations, along those of state-of-the-art metaheuristics solvers on classical hardware.

1 Introduction

The Costas Array Problem is an interesting hard combinatorial problem that has been tackled for many years but for which no completely satisfactory solving method yet exists.

Historically, Costas arrays were developed in 1965 by John P. Costas to describe a novel frequency hopping pattern for SONARs with optimal autocorrelation properties in order to avoid noise [12]. Costas Arrays also have applications in radars, wireless telecommunications and cryptography and research is still active in using methods based on Costas Arrays for the detection of closely spaced targets in radar applications [11] or synthetic aperture radar (SAR) imagery [52].

Interestingly, the Costas Array Problem (CAP) is difficult to solve but easy to describe: A Costas array of size n is an $n \times n$ grid containing n marks such that there is exactly one mark per row and per column and the $n(n-1)/2$ vectors joining the marks are all different. It can be viewed as a permutation problem of size n, where the value j of the ith element of the permutation represents the column j on which the mark on row i is set.

A very complete survey on Costas arrays can be found in [16]. Finding a Costas array of size n is in NP and the enumeration of all Costas array of size n is NP-complete. In the 1980s, several algorithms were proposed to build a Costas

L. Pérez Cáceres and S. Verel (Eds.): EvoCOP 2022, LNCS 13222, pp. 143–158, 2022.
https://doi.org/10.1007/978-3-031-04148-8_10

array for a given n, such as the Welch construction and the Golomb construction, see for instance [44]. These methods work up to $n = 29$ but cannot built Costas arrays of size 32 and some higher non-prime sizes. Nowadays, after many decades of research, it remains unknown whether there exist any Costas arrays of size 32 or 33. Another difficult problem is to enumerate all Costas arrays for a given size. Moreover, if the number of solutions for a given size instance n increases from $n = 1$ to $n = 16$, it then decreases from $n = 17$ onward, reaching 164 for $n = 29$ [18]. Only a few solutions (and not all) are known for $n = 30$ and $n = 31$. It has indeed been proved recently, after a few decades of conjecture, that CAP solutions are in exponential decay as n growing, making thus CAP a very difficult problem for higher values of n [10]. Local search and metaheuristics methods have been used for solving CAP, and the best of such methods, an iterated local search metaheuristics [6,14], will find a solution for CAP with $n = 18$ in a few seconds and $n = 20$ in a few minutes on a single-core computer. Parallel versions of this method, either with independent [6] or cooperative [38] parallelism, can significantly improve performance.

With the emergence of quantum computers in the last decade and in particular the use of systems based on quantum annealing for solving combinatorial problems, an interesting question is to know whether CAP could be solved efficiently by quantum computing. Quantum Annealing (QA) has been proposed two decades ago by [30] and [19] and incorporated one decade ago as the core computational mechanism in the D-Wave machines [5,28], creating thus a whole new domain for solving optimization problems. Compared to the more well-known *gate-based* model in quantum computing, QA is an alternative type of computation that can be effective in the current NISQ (Noisy Intermediate Scale Quantum) era. QA is derived from simulated annealing [31], but taking advantage of the *quantum tunneling* effect to overcome energy barriers and therefore escape local minima during the computation. Moreover, QA is related to the QUBO formalism (Quadratic Unconstrained Binary Optimization) in combinatorial optimization [32]. QUBO has thus become a de facto input language for quantum computers such as D-Wave systems or "quantum-inspired" dedicated CMOS hardware such as Fujitsu's Digital Annealer Unit (DAU) [2]. Therefore an interesting issue is to know if modeling problems in QUBO and executing them on QA hardware is opening a new field for combinatorial optimization that could be competitive with the best classical algorithms existing for various combinatorial problems. However, whereas several papers [34,41] show the performance of quantum annealing systems on graph-based combinatorial optimization benchmarks which can be easily modeled as QUBO such as Max-cut, TSP or MIS and compare then to classical solvers (e.g. Gurobi or CPLEX), there is a lack of performance analysis for more complex constrained optimization problems and a lack of comparison with state-of-the-art metaheuristics methods, i.e. not only basic simulated annealing.

We would like to propose the Costas Array Problem (CAP) as a benchmark in order to evaluate quantum annealing as an efficient alternative way to solve combinatorial optimization problems with respect to classical methods, in particular state-of-the-art local search metaheuristics. This problem is coming from a real-life application, is easy to understand and can be modeled in QUBO, but it is difficult to solve to optimality because it involves many constraints that could interfere in the annealing process. It therefore seems a good benchmark to further test the use of quantum annealing solvers for complex combinatorial optimization and assess their performance.

This paper is organized as follows. Section 2 presents the QUBO format, its relation with the Ising model and the basic notions of quantum annealing. Section 3 details how to encode constraints in QUBO, including rather complex ones such as the `all-different` constraint, while Sect. 4 describes how to model CAP in QUBO and indeed how to perform the transformation from Higher Order Binary Optimization (HOBO) to QUBO. We finally present in Sect. 5 the results obtained by implementing the QUBO models on the quantum D-Wave systems and the quantum-inspired Fujitsu Digital Annealing Unit and Fixstars Amplify Annealing Engine. A short conclusion ends the paper.

2 QUBO and Quantum Annealing

The QUBO paradigm is conceptually very simple but has shown to be quite powerful at modeling various types of combinatorial problems. Moreover, this paradigm is indeed very similar to the Ising model which is used in statistical physics, making it possible to envision a direct computation with hardware systems based on quantum annealing.

It has been shown in [33] that classical NP problems can be cast in the framework of the Ising model and therefore modeled as QUBO. Moreover, according to [22], many combinatorial optimization problems can be modeled directly in QUBO: graph problems (graph coloring, Max-Cut, Vertex Cover, Max-Clique, Maximum Independence Set), Traveling Salesman Problems, Assignment Problems, Task Allocation Problems, Knapsack Problems, and also simple cases of SAT and Constraint Satisfaction Problems (CSPs).

2.1 QUBO

A Quadratic Unconstrained Binary Optimization (QUBO) problem consists in minimizing or maximizing a quadratic expression over a finite number of Boolean variables.

Consider n Boolean variables $x_1, ..., x_n$ and a quadratic expression over $x_1, ..., x_n$ to minimize:

$$\sum_{i=1}^{n} q_i x_i + \sum_{i=1}^{n} q_i' x_i^2 + \sum_{i,j} q_{ij} x_i x_j$$

As $x_i \in \{0, 1\}$, $x_i = x_i^2$ and thus the first two terms can be merged and integrated in the third one. Thanks to commutativity, this expression can therefore be further simplified to:

$$\sum_{i \leq j} q_{ij}' x_i x_j$$

It is therefore usual to represent a QUBO problem by a vector x of n binary decision variables and a square $n \times n$ matrix Q with coefficients q_{ij}, which can be given in symmetric or in upper triangular form, without loss of generality. The QUBO problem can thus be written as: $minimize/maximize \quad y = x^t Q x$ where x^t is the transpose of x

The reader is referred to [32] for an early survey of QUBO and to [22] for a recent and didactic presentation.

2.2   The Ising Model

Much of the recent interest for QUBO lies in the fact that it is very close to Ising model which has been used in physics for many years. The Ising model is a mathematical model of ferromagnetism in statistical mechanics. It consists of discrete variables that represent magnetic dipole moments of atomic "spins" that can be in one of two states ($+1$ or -1). The spins are arranged in a graph, usually a lattice, allowing each spin to interact with its neighbors. In the Ising model, the Hamiltonian (energy function) is traditionally defined as:

$$H = -\sum_i h_i \sigma_i - \sum_{i,j} J_{ij} \sigma_i \sigma_j$$

where J_{ij} is the coupling strength between spins i and j, and h_i is the bias on spin i.

This is exactly like a QUBO problem, as spin variables σ_i can be exchanged with Boolean variables x_i as follows: $\sigma_i = 2x_i - 1$. Optimization problems modeled as QUBO can therefore be straightforwardly converted to Ising Hamiltonians, the ground states of which correspond to the minimal solutions of the QUBO problem. These ground states can be computed by devices using adiabatic quantum evolution, an example of which being quantum annealing [37].

2.3   Quantum Annealing

There are currently two main paradigms for quantum computing, the *gate model*, in which computation is realized by a circuit with "quantum gates" which are unitary operations (e.g. Hadamard gate, Toffoli gate, etc.) analogous to logic gates in classical computers, and the *quantum adiabatic computing* paradigm in which problems are encoded in quantum Hamiltonians (energy functions) and quantum dynamics is used to find solutions (ground states of minimal energy). The first paradigm is exemplified by the quantum hardware systems developed by IBM, Google, Microsoft, Rigetti, etc., while the second one is exemplified by the quantum hardware systems based on quantum annealing produced by D-Wave Systems Inc. Quantum computers operates on quantum bits or *qubits*, and the gate-based quantum computers are currently limited to about a hundred qubits (e.g. 127 for the IBM machine, 54 for the Google machine), with IBM aiming at a system with 1000 qubits in 2023. On the other hand, the D-Wave Advantage system (2020) has more than 5600 qubits.

Quantum Annealing (QA) has been proposed by [30] and [19] and incorporated as the core computational mechanism in the D-Wave computers, in which qubits are implemented by superconducting Josephson junctions. Another implementation approach is the photonics-based Coherent Ising Machine [26] developed by NTT. A detailed description of the physics of QA is out of the scope of this paper, and we refer the reader to [48] for a full treatment and to [24] for a survey on current issues and perspectives regarding the implementation of physical devices. QA thus essentially performs an adiabatic evolution to optimize some function of interest which is encoded as a *transverse field* Hamiltonian. The basic idea is to start the computation by placing the initial system qubits in a ground state that is easy to prepare and to slowly modify the energy landscape by adding the transverse Hamiltonian corresponding to the problem to be solved while fading away the original Hamiltonian. The *Quantum Adiabatic Theorem* states that, under a sufficiently slow evolution of the system, it is likely to remain in a ground state, therefore ensuring that the final state will very likely encode the solution of the problem represented in the transverse Hamiltonian.

From a metaheuristic viewpoint, QA can be seen as derived from simulated annealing but with a different manner to escape local minima. Indeed, if in simulated annealing escaping from local minima is done by accepting with a certain probability non-decreasing moves, this is performed in QA by the phenomenon of *quantum tunneling* which makes it possible to traverse energy potential barriers in the energy landscape as long as they are not too large, i.e. high and narrow peaks do not cause a problem. This quantum tunneling effect is physically performed in quantum hardware based on QA computations.

2.4 Quantum-Inspired Annealing

In addition to quantum annealing systems such as D-Wave computers, several *quantum-inspired* annealing hardware have been developed in the last few years, such as for instance Fujitsu's Digital Annealer Unit (DAU) which is a special-purpose CMOS hardware implementing in its latest version (DA3) 100,000 bits connected by a complete graph and associated couplers. [2] details the architecture and the dedicated annealing algorithm that incorporates parallel tempering and parallel-trial Monte Carlo updates, as well as offset addition to simulate quantum tunneling. "Quantum-inspired" (a.k.a. "digital") annealing thus consists in simulating quantum annealing with classical electronics, that could be either dedicated hardware (e.g., Fujitsu's DAU and Hitachi's CMOS Annealing Machine) or clusters of parallel general-purpose hardware (e.g., Fixstars Amplify Annealing Engine, NEC's Annealer and Toshiba's Simulated Bifurcation Machine).

Fixstars Amplify Annealing Engine (AE) [35] is a Graphics Processing Unit (GPU)-based annealing with a capacity of 65,536 bits (i.e. Boolean variables). Toshiba's Simulated Bifurcation Machine (SBM) [23] is based on software simulation of adiabatic and chaotic (ergodic) evolution of nonlinear Hamiltonian systems on general-purpose parallel hardware and can supposedly handle up to 1 million bits, while an hardware FPGA version has a capacity of 8000 bits

[49]. Hitachi's CMOS Annealing Machine is a special ASIC/FPGA hardware unit [53], and several units combined together have a capacity of 150,000 bits connected by a King's graph [47]. NEC has developed a software-based digital annealer running on their own SX-AuroraTSUBASA Vector Engine with a capacity equivalent to 100,000 fully connected bits [15].

All these systems use QUBO as an input language.

3   Constrained Models in QUBO

The classical combinatorial optimization problems usually modeled in Ising or QUBO models [22,33] are in general relatively simple and involve basic constraints that can be introduced by modifying the objective function. Indeed, although the "U" in QUBO stands for "Unconstrained", it is easy to introduce constraint expressions in QUBO models as *penalties* in the objective function to minimize, that is, as quadratic expressions whose value is minimal when the constraint is satisfied. An easy way to formulate such a penalty is to create a quadratic expression which has value 0 if the constraint is satisfied and a positive value if the constraint is not satisfied, representing somehow the degree of violation of the constraint. This is indeed the manner in which constraints have been represented in constraint-based local search for many years [9,20,25]. When several constraints c_i have to be integrated in the QUBO problem, the corresponding penalties should be added to the objective function to optimize with penalty coefficients p_i large enough to make such constraints "hard", whereas the objective function is to be considered "soft". Finding the best penalty coefficients for integrating various constraints in the objective function can be a difficult task, see [46] for a discussion on a concrete example. A set of penalty expressions for basic pseudo-Boolean constraints on 2 or 3 Boolean variables that can be used for modeling constraint problems in QUBO can be found in [22], e.g. the penalty for a constraint $x \leq y$ will be $x - xy$ and for $x + y = 1$ it will be $1 - x - y + 2xy$.

More complex constraints can also be modeled. For instance, the so-called "One-Hot" constraint in QA literature [40], also known as the `exactly-one` constraint in the Constraint Programming literature, ensures that among a list of Boolean variables $(x_1, \ldots, x_n)$ only one variable has value 1 while others have value 0. This constraint appears very frequently when modeling combinatorial problems and to encode this constraint in a QUBO model we need the pseudo-boolean constraint:

$$\sum_{i=1}^{n} x_i = 1 \qquad \Longleftrightarrow \qquad (\sum_{i=1}^{n} x_i - 1)^2 = 0$$

The corresponding penalty to add to the QUBO objective function is thus:

$$-\sum_{i=1}^{n} x_i + 2 \sum_{i<j} x_i x_j$$

Building on the "One-Hot" constraint, one can define the QUBO penalty for the constraint known as "two-way One-Hot" in the QA community, which corresponds to the `all-different` constraint on integer variables [21], and has proved to be very useful for modeling many problems in the domain of Constraint Programming. This constraint is the archetypal model of a so-called *global* constraint[1] and will be needed in modeling the Costas Array Problem.

For n variables $x_1, \ldots, x_n$ with values taken in the domain $\{1, \ldots, n\}$, the constraint `all-different`$(x_1, \cdots, x_n)$ will ensure that each variable has a different value, i.e. that $(x_1, \ldots, x_n)$ is a permutation of $(1, \ldots, n)$.

To translate this constraint in QUBO, we need n^2 Boolean variables x_{ij} that have value 1 if the original variable x_i has j and value 0 otherwise. `all-different`$(x_1, \ldots, x_n)$ is then encoded by the $2 \times n$ One-Hot constraints:

$$\forall i \in \{1, n\} \ \sum_{j=1}^{n} x_{ij} = 1 \qquad \forall j \in \{1, n\} \ \sum_{i=1}^{n} x_{ij} = 1$$

As explained above, each One-Hot constraint leads to a quadratic penalty expression, and adding everything together gives after simplification the following penalty for a QUBO model:

$$- \sum_{i=1, j=1}^{n} x_{ij} + \sum_{k=1}^{n} \sum_{i<j} x_{ki} x_{kj} + \sum_{k=1}^{n} \sum_{i<j} x_{ik} x_{jk} \tag{1}$$

Observe that the term $\sum_{i=1, j=1}^{n} x_{ij}$ is needed and should appear negatively in the penalty in order to maximize the number of Boolean to be set to 1 in the solution. The two other terms ensure that each variable has a unique value and that each value is given to a unique variable.

4   The Costas Array Problem

A Costas array is an $n \times n$ grid containing n marks such that there is exactly one mark per row and per column and the $n(n-1)/2$ vectors joining the marks are all different. Figure 1 depicts an example of a Costas array of size 5.

In addition to the constructive methods such as the Welch construction and the Golomb construction [44], which do not work for all value of n and cannot generate all costas arrays for a given n, several approaches in the domain of combinatorial optimization have been used for solving CAP: iterative local search, ant colony optimization and constraint programming.

Indeed, the Costas Array Problem (CAP) has been proposed as a combinatorial problem benchmark for the first time more than 30 years ago in [50], where it was solved (for small values of n) by Logic Programming and Constraint Programming. An improved model in Constraint Programming was later proposed

---

[1] Let us note that recently [3] proposed some "quantum-accelerated filtering algorithms" to encode the `all-different` constraint in the *gate model* of quantum computing, which is very different from what we consider here (QUBO and QA).

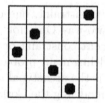

Fig. 1. A solution to the Costas Array Problem of size 5

by Barry O'Sullivan in the MiniZinc Language [39] and various search strategies have been presented in [45]. CAP has also been modeled and solved by local search metaheuristics: first by Dialectic Search [29], and then more efficiently by Adaptive Search, an iterative local search method [6,14]. Performance-wise, local search is clearly faster than constraint programming.

Let us briefly present the basic ideas of these models. It is convenient to see the CAP as a permutation problem where each permutation represents a configuration of n marks on the grid. The solution to the CAP of size 5 depicted in Fig. 1 can thus be represented by the permutation $(3, 4, 2, 1, 5)$. In Constraint Programming or local search, CAP can be modeled by considering n variables $(x_1, \ldots, x_n)$ with domains $\{1, \cdots, n\}$ which forms a permutation of $(1, 2, \ldots, n)$, i.e. with an implicit `all-different` constraint over the variables x_i. A variable x_i has value j iff there is a mark at column i and row j. To take into account constraints on vectors between marks (which must all be different) it is convenient to use the so-called *difference triangle* [17]. This triangle contains $n - 1$ rows, each row corresponding to a distance d. The dth row of the triangle is of size $n - d$ and contains the differences between any two marks at a distance d, i.e. the values $x_{i+d} - x_i$ for all $i = 1, \ldots, n - d$. Ensuring that all vectors are different comes down to ensure the triangle contains no repeated values on any given row.

Table 1. Performance of various methods on CAP (timings in seconds)

n	Iterated Local Search	Ant Colony Optimization	Constraint Programming	
	Adaptive Search version 1.0.5	mDRACO [51]	MiniZinc + Gecode	OR-Tools (Python API)
12	0.00	0.11	2.89	0.01
17	0.25	17.75	254	794
18	1.95	95.99	23.9	54.4
19	18.8	798.7	10510	28934
20	214.5	-	-	-

We do not intend in this paper to perform a detailed comparison of those methods for solving CAP, but as a matter of reference, we summarize their performance in the Table 1. The timings for Adaptive Search and Constraint Programming[2] where done by running the solvers on a Dell computer equipped with a i7-8665U CPU at 1.90 GHz, while the performances of mDRACO are taken from [51]. All timings are in seconds.

4.1 Modeling the Costas Array Problem in HOBO

CAP is in fact rather simple to model as HOBO (Higher-Order Binary Optimization), with an objective function using quartic terms (i.e., of order 4). We will present this basic model and then detail how it can be quadratized in order to provide a QUBO model.

To model CAP, we can consider Boolean variables x_{ij} with value 1 if there is a mark on the cell on column i and row j and 0 otherwise. The basic constraint that we have n marks that indeed represent a permutation of $(1, 2, \ldots, n)$ amounts to an `all-different`/"two-way One-Hot" constraint between the x_{ij}, and thus to a penalty given by Eq. 1 in Sect. 3.

The property that all the vectors linking two marks are all different can be represented by a pseudo-Boolean formula involving the variables x_{ij}:

$$\sum_{\substack{1 \leq i < i' < n,\ j,j' \in \{1, \cdots, n\},\ k,l \in \{-n+1, \cdots, n-1\} \\ 1 < i+k \leq n-2,\ 1 \leq j+l \leq n,\ 2 < i'+k \leq n,\ 1 < j'+l \leq n}} x_{ij}\, x_{i+k,j+l}\, x_{i'j'}\, x_{i'+k,j'+l} = 0 \quad (2)$$

This states that if we have a mark on cell x_{ij} and on cell $x_{i+k,j+l}$ (observe that l can be negative), then for every cell $x_{i'j'}$ on a column i' greater than i we cannot have a mark on a cell $x_{i'+k,j'+l}$, i.e., at the same distance in rows (k) and columns (l) from $x_{i',j'}$.

The objective function to minimize in the HOBO model is thus given by the sum of the left-hand side of Eq. 2 and the penalty expression for the `all-different`/"two-way One-Hot" constraint:

$$\sum_{\substack{1 \leq i < i' < n, \\ j,j' \in \{1, \cdots, n\}, \\ k,l \in \{-n+1, \ldots, n-1\} \\ 1 < i+k \leq n-2, \\ 1 \leq j+l \leq n, \\ 2 < i'+k \leq n, \\ 1 < j'+l \leq n}} x_{ij}\, x_{i+k,j+l}\, x_{i'j'}\, x_{i'+k,j'+l} - \sum_{i=1,j=1}^{n} x_{ij} + \sum_{k=1}^{n}\sum_{i<j} x_{ki}x_{kj} + \sum_{k=1}^{n}\sum_{i<j} x_{ik}x_{jk}$$

A solution for CAP is found if this expression can be minimized to zero.

This is however not a quadratic but a *quartic* expression on Boolean variables. Therefore to model the Costas Array Problem, we need to reduce the higher-degree polynomial into a quadratic one, i.e. transform a Higher-Order Binary Optimization (HOBO) problem into a QUBO problem. It is known for a few decades that every pseudo-Boolean function admits a quadratization [43], and this transformation is performed by introducing new ancillary variables.

---

[2] We used two versions derived from Barry O'Sullivan's model by Hakan Kjellerstrand, available at http://www.hakank.org/minizinc: one written in MiniZinc with Gecode as back-end solver and one written in Python for the Google OR-Tools solver.

4.2   Modeling the Costas Array Problem in QUBO

The *quadratization* of a pseudo-Boolean function $f(x)$ on a vector of variables x is a quadratic function $g(x, y)$ on the original variables x and ancillary variables y that satisfies $f(x) = min\{g(x,y)|y \in \{0,1\}^m, x \in \{0,1\}^n\}$ [1]. Although simple techniques for quadratization exist for a long time, it is desirable to produce a quadratization with a small number of auxiliary variables, see [13] for an extensive survey of various techniques. For instance, the classic Rosenberg technique [43] will reduce a monomial of degree n by introducing $n - 2$ auxiliary variables, while Ishikawa technique [27] will only introduce $\lfloor (n - 1)/2 \rfloor$ auxiliary variables. Recently, [4,42] extended Ishikawa technique and proposed a method that will only introduce $\lceil log(n) \rceil - 1$ auxiliary variables for a positive monomial. Both techniques would be equivalent for a quartic positive monomial, adding one extra-variable and using the formula [27]:

$$xyzt = min_{w \in \{0,1\}} (w(-2(x + y + z + t) + 3) + xy + yz + zx + tx + ty + tz)$$

As a single ancillary variable is created for each quartic monomial, we need to count the total number of quartic monomial in order to count the total number of ancillary variables. Looking again at Eq. 2, we can see that one monomial is created for each triplet of marks $(x_{i_1 j_1}, x_{i_2 j_2}, x_{i_3 j_3})$. The forth mark involved in the monomial is uniquely determined in order to form a parallelogram together with $(x_{i_1 j_2}, x_{i_2 j_2}, x_{i_3 j_3})$. To cover all possible triplets of marks, we have to choose 3 different numbers in $\{1, n\}$, thus $\binom{n}{3} = \frac{n(n-1)(n-2)}{6}$ possibilities, for the row indices and 3 different numbers in $\{1, n\}$, thus again $\binom{n}{3}$ possibilities, for the column indices. Thus the total number of quartic monomials is equal to:

$$\frac{n(n-1)(n-2)}{6} \times \frac{n(n-1)(n-2)}{6} = \frac{n^2(n-1)^2(n-2)^2}{36}$$

The number of extra variables to be introduced for the quadratization of the objective function is therefore in $\mathcal{O}(n^6)$, which could be quite a large number as n grows. For instance, the QUBO model of a small CAP of size 12 will have nearly 50,000 variables.

Another way to perform quadratization is to use *pairwise covers* [1], which seems to work well in practice. The basic idea of pairwise cover is to introduce new boolean variables that represent pairs of initial Boolean variables appearing in non-quadratic monomials. In the QUBO model for CAP described above, this consists in introducing new variables d_{ij}^{kl}, $\forall i, j, k, l \in \{1, \cdots, n\}$ and $i < k$, which have value 1 if and only if there is one mark on column i and row j and one mark on column k and row l. Therefore a variable d_{ij}^{kl} represents the pair (x_{ij}, x_{kl}).

The Costas Array property can be easily rephrased in term of d_{ij}^{kl}: no two variables should have the same difference between indices ($k-l$ and $l-j$), that is:

$$\forall d_{ij}^{kl}, \forall d_{i'j'}^{k'l'} : k - i \neq k' - i' \text{ or } l - j \neq l' - j'$$

This leads to the following quadratic pseudo-Boolean constraint:

$$\sum d_{ij}^{kl} \, d_{ab}^{a+k-i,b+j-l} = 0$$

Therefore the corresponding penalty for the QUBO model is;

$$\sum_{\{i,j,k,l,a,b\in\{1,\cdots,n\},i<k,i<a,a+k-i\leq n,b+j-l\leq n\}} \sum d_{ij}^{kl} \, d_{ab}^{a+k-i,b+j-l}$$

Considering the number of auxiliary variables introduced, the pairwise cover technique will introduce $\frac{n^2(n^2-1)}{2}$ extra variables, i.e. $\mathcal{O}(n^4)$, while the Ishikawa technique would introduce $\frac{n^2(n-1)^2(n-2)^2}{36}$ extra variables, i.e. $\mathcal{O}(n^6)$. In the QUBO models obtained (including the n^2 initial variables x_{ij}) for CAP with n = 10, this means 4600 variables versus 14,500 variables and, with n = 12, 9648 variables versus 48,544.

As the maximal number of QUBO variables is limited in quantum and quantum-inspired solvers, it is therefore better to use the pairwise technique. Observe anyway that, even in the case of the pairwise cover technique, the CAP model for $n = 20$ would have 76,400 variables due to the large number of variables generated by the transformation from HOBO to QUBO, which is close to the limit of 100,000 variables of the largest systems today.

5   Implementation Results

We implemented the CAP model using pairwise cover for quadratization in order to have a manageable number of variables, and we give in this section the preliminary performance results for this implementation on quantum and quantum-inspired annealers. Better results could probably be achieved with some extra tuning in the program (e.g. for the penalty coefficients) or in the solver (e.g. temperature control for the annealing process), but would probably not change qualitatively the conclusions of the experiment.

5.1   Implementation on D-Wave

The latest D-Wave Advantage system features about 5600 qubits and both a quantum processing unit (QPU) solver and hybrid solvers mixing classical software decomposing the original problem in smaller sub-problems and calls to the QPU solver. Hybrid solvers make it possible to tackle larger problems, but at the price of performance. D-Wave systems also include a classical software-based Tabu Search QUBO solver (`QBSolv`).

However, the maximal number of QUBO variables (represented by qubits) available for direct execution on QPU is in general quite lower than 5600 because D-Wave systems do not implement a fully connected graph between qubits[3], and therefore some qubits are also needed to encode extra connections ("couplers")

---

[3] More precisely, the Chimera architecture of the D-Wave 2000X computer has only a 6-way qubit connectivity (meaning that each qubit is physically connected to at most 6 other qubits) and the Pegasus architecture of the D-Wave Advantage has a 15-way qubit connectivity.

between qubits. According to [36], the maximal fully connected graph that can be simulated on the D-Wave Advantage is limited to 177 logical qubits (with maximal chain lengths of 17). This drastically reduces the problem size for which the QA hardware could be used when full connectivity is needed because of global constraints. This process of mapping a QUBO problem graph onto a subgraph of the hardware graph is called *minor embedding* [7,8] and has received much attention in the literature related to the D-Wave systems. Minor embedding use additional qubits for simulating the missing hardware connections. One key idea is to create chains of *physical qubits* that will represent a single *logical qubit* (i.e. a QUBO variable) thus allowing more potential connections for this qubit. As missing connections have to be simulated by using extra qubits, there is therefore a difference between the number of logical qubits (number of Boolean variables of the QUBO model) and the number of physical qubits (number of qubits used in the hardware graph) needed to encode the QUBO model in the hardware QA, depending on the connectivity needed.

Unfortunately, even for CAP of small size (e.g. $n = 8$), D-Wave transformation software is not able to find a minor embedding in order to run the problem on the QPU solver. This is probably because of the large number of constraints, requiring many connections between qubits to be active. Let us note that for CAP of size $n = 8$, QBSolv and Hybrid Solver can achieve an optimal solution satisfying the Costas property: on average in 4.9s for Hybrid Solver and 8.5s for QBSolv. However for $n \geq 10$ both solvers cannot find an optimal solution (with zero conflicts, i.e. satisfying all Costas property constraints) within a timeout of one minute.

5.2   Implementation on Fixstars Amplify AE and Fujitsu DA3

We give in this section the performance results for quantum-inspired digital annealers. As these systems implement (in hardware or software) the complete connection graph between all QUBO variables and can cope with 65,536 QUBO variables (Fixstars Amplify AE) or 100,000 QUBO variables (Fujitsu Digital Annealer version 3), they can solve larger instances of CAP. Table 2 details the performance of Fixstars Amplify AE and Fujitsu DA3 on the CAP; timings are the average of 10 runs and are in seconds. We fixed a timeout of one minute before stopping the solvers (which can therefore reach an optimal solution or not)[4]. For n = 8, both solvers can solve the problem to optimality before the timeout for all runs. For n = 10, Fixstars Amplify AE cannot solve for all runs the problem to optimality within the time limit of 1 minute and has a success rate of 30%, while Fujitsu DA3 has a success rate of 100%. For n = 12, both solvers cannot solve the problem to optimality for all runs; Fixstars Amplify AE cannot reach any optimal solution within 1 minute, while Fujitsu DA3 has a success rate of 20%. Table 2 summarizes for each CAP instance the success rate (within 1 minute timeout), the average time to reach the optimal solution

---

[4] There is a limit of 60s per job when using Fixstars Amplify AE with developer accounts, we therefore took the same time limit for Fujitsu's DAU.

(if any is achieved) and the average time to reach a quasi-optimal solution equal to optimal solution + 0.1% (when the optimal cannot be reached)

Table 2. Quantum-inspired annealers performance on CAP (timings in seconds)

n	Fixstars AE			Fujitsu DA3		
	success rate (timeout = 1 mn)	time to solution	time to opt. +0.1%	success rate (timeout = 1 mn)	time to solution	time to opt. + 0.1%
8	100%	0.084	-	100%	0.424	-
10	30%	18.9	16.7	100%	8.89	-
12	0%	-	18.0	20%	37.7	31.1

As expected, quantum-inspired solvers can solve larger CAP instances than D-Wave solvers and have better performance on this problem than D-Wave Hybrid Solver as they can solve CAP for $n = 8$ within a few hundreds of milliseconds rather than a few seconds. However, their performance is far below that of metaheuristics or constraint programming, which can solve CAP for $n = 12$ in a few milliseconds and CAP for $n = 18$ in a few seconds, cf. Table 1 in Sect. 4.

6 Conclusion

We have detailed how to model the Costas Array Problem (CAP) in QUBO in order to solve it by quantum annealing on quantum systems and quantum-inspired hardware. QUBO, which is the input language for quantum computers based on the quantum annealing paradigm, is a simple but expressive way to model various combinatorial optimization problems, and it is indeed possible to encode rather complex constraints, such as the ones needed for CAP, in QUBO. However, for CAP the translation from HOBO (Higher-Order Binary Optimization) to QUBO models generates a lot of ancillary Boolean variables that prevent solving large-scale models on current quantum annealing hardware. Quantum-inspired annealers such as Fujitsu Digital Annealer (DA3) or Fixstars Annealer Engine (AE) perform better, but are still limited to small instances, e.g. up to $n = 12$.

From our experiments, it seems that the current generation of quantum and quantum-inspired hardware is not mature enough for handling even medium size constraint-based satisfaction or optimization problems involving complex constraints. They are more suited to pure optimization problems or optimization problems with simple constraints. This experiment with CAP is nevertheless a proof-of-concept that quantum annealing can be used for solving difficult constrained problems and, as quantum and quantum-inspired hardware will improve in the future years, they might become very competitive with respect to state-of-the-art solvers using metaheuristics methods on classical hardware.

Indeed, if quantum annealers or quantum-inspired hardware annealer could extend by one or two orders of magnitude the number of available bits/qubits with respect to current systems (maybe in 5 to 10 years?), this would probably be challenging for the best solvers on classical hardware.

References

1. Anthony, M., Boros, E., Crama, Y., Gruber, A.: Quadratic reformulations of nonlinear binary optimization problems. Math. Program. **162**(1–2), 115–144 (2017)
2. Aramon, M., Rosenberg, G., Valiante, E., Miyazawa, T., Tamura, H., Katzgraber, H.G.: Physics-inspired optimization for quadratic unconstrained problems using a digital annealer. Front. Phys. **7**, 48 (2019)
3. Booth, K.E.C., O'Gorman, B., Marshall, J., Hadfield, S., Rieffel, E.: Quantum-accelerated global constraint filtering. In: Simonis, H. (ed.) CP 2020. LNCS, vol. 12333, pp. 72–89. Springer, Cham (2020). https://doi.org/10.1007/978-3-030-58475-7_5
4. Boros, E., Crama, Y., Rodríguez-Heck, E.: Compact quadratizations for pseudo-boolean functions. J. Comb. Optim. **39**(3), 687–707 (2020)
5. Bunyk, P.I., et al.: Architectural considerations in the design of a superconducting quantum annealing processor. IEEE Trans. Appl. Supercond. **24**(4), 1–10 (2014)
6. Caniou, Y., Codognet, P., Richoux, F., Diaz, D., Abreu, S.: Large-scale parallelism for constraint-based local search: the costas array case study. Constraints **20**(1), 30–56 (2014). https://doi.org/10.1007/s10601-014-9168-4
7. Choi, V.: Minor-embedding in adiabatic quantum computation: I. the parameter setting problem. Quantum Inf. Process. **7**(5), 193–209 (2008)
8. Choi, V.: Minor-embedding in adiabatic quantum computation: II. minor-universal graph design. Quantum Inf. Process. **10**(3), 343–353 (2011)
9. Codognet, P., Diaz, D.: Yet another local search method for constraint solving. In: Steinhöfel, K. (ed.) SAGA 2001. LNCS, vol. 2264, pp. 73–90. Springer, Heidelberg (2001). https://doi.org/10.1007/3-540-45322-9_5
10. Correll, B.: The density of costas arrays and three-free permutations. In: 2012 IEEE Statistical Signal Processing Workshop (SSP), pp. 492–495 (2012)
11. Correll, B., Beard, J.K., Swanson, C.N.: Costas array waveforms for closely spaced target detection. IEEE Trans. Aerosp. Electron. Syst. **56**(2), 1045–1076 (2020)
12. Costas, J.: A study of detection waveforms having nearly ideal range-doppler ambiguity properties. Proc. IEEE **72**(8), 996–1009 (1984)
13. Dattani, N.: Quadratization in discrete optimization and quantum mechanics (2019)
14. Diaz, D., Richoux, F., Codognet, P., Caniou, Y., Abreu, S.: Constraint-based local search for the costas array problem. In: Hamadi, Y., Schoenauer, M. (eds.) LION 2012. LNCS, pp. 378–383. Springer, Heidelberg (2012). https://doi.org/10.1007/978-3-642-34413-8_31
15. Dote, S.: NEC's initiative in quantum computing. Presented at D-Wave Qubits Conference 2021, October 2021 (2021)
16. Drakakis, K.: A review of costas arrays. J. Appl. Math. **2006**, 1–32 (2006)
17. Drakakis, K., Gow, R., Rickard, S.: Distance vectors in costas arrays. In: Proceedings of CISS 2008, 42nd Annual Conference on Information Sciences and Systems, pp. 1234–1239. IEEE Press (2008)
18. Drakakis, K., Iorio, F., Rickard, S., Walsh, J.: Results of the enumeration of costas arrays of order 29. Adv. Math. Commun. **5**(3), 547–553 (2011)

19. Farhi, E., Goldstone, J., Gutmann, S., Lapan, J., Lundgren, A., Preda, D.: A quantum adiabatic evolution algorithm applied to random instances of an np-complete problem. Science **292**(5516), 472–475 (2001)
20. Galinier, P., Hao, J.: A general approach for constraint solving by local search. J. Math. Model. Algorithms **3**(1), 73–88 (2004)
21. Gent, I.P., Miguel, I., Nightingale, P.: Generalised arc consistency for the alldifferent constraint: an empirical survey. Artif. Intell. **172**(18), 1973–2000 (2008)
22. Glover, F.W., Kochenberger, G.A., Du, Y.: Quantum bridge analytics I: a tutorial on formulating and using QUBO models. 4OR **17**(4), 335–371 (2019)
23. Goto, H., Tatsumura, K., Dixon, A.R.: Combinatorial optimization by simulating adiabatic bifurcations in nonlinear hamiltonian systems. Sci. Adv. **5**(4), eaav2372 (2019)
24. Hauke, P., Katzgraber, H.G., Lechner, W., Nishimori, H., Oliver, W.D.: Perspectives of quantum annealing: methods and implementations. Rep. Prog. Phys. **83**(5), 054401 (2020)
25. Hentenryck, P.V., Michel, L.: Constraint-Based Local Search. MIT Press, Cambridge (2005)
26. Inagaki, T., et al.: A coherent ising machine for 2000-node optimization problems. Science **354**(6312), 603–606 (2016)
27. Ishikawa, H.: Transformation of general binary MRF minimization to the first-order case. IEEE Trans. Pattern Anal. Mach. Intell. **33**(6), 1234–1249 (2011)
28. Johnson, M., et al.: Quantum annealing with manufactured spins. Nature **473**, 194–198 (2011)
29. Kadioglu, S., Sellmann, M.: Dialectic search. In: Gent, I.P. (ed.) CP 2009. LNCS, vol. 5732, pp. 486–500. Springer, Heidelberg (2009). https://doi.org/10.1007/978-3-642-04244-7_39
30. Kadowaki, T., Nishimori, H.: Quantum annealing in the transverse ising model. Phys. Rev. E **58**, 5355–5363 (1998)
31. Kirkpatrick, S., Gelatt, C.D., Vecchi, M.P.: Optimization by simulated annealing. Science **220**(4598), 671–680 (1983)
32. Kochenberger, G., et al.: The unconstrained binary quadratic programming problem: a survey. J. Comb. Optim. **28**(1), 58–81 (2014). https://doi.org/10.1007/s10878-014-9734-0
33. Lucas, A.: Ising formulations of many np problems. Front. Phys. **2**, 5 (2014)
34. Matsubara, S., et al.: Digital annealer for high-speed solving of combinatorial optimization problems and its applications. In: 2020 25th Asia and South Pacific Design Automation Conference (ASP-DAC), pp. 667–672 (2020)
35. Matsuda, Y.: Research and development of common software platform for ising machines. In: 2020 IEICE General Conference (2020). https://amplify.fixstars.com/docs/_static/paper.pdf
36. McGeoch, C., Farré, P.: The advantage system: performance update, Technical report, D-Wave, 01 October 2021
37. McGeoch, C.C.: Adiabatic Quantum Computation and Quantum Annealing: Theory and Practice. Morgan & Claypool, San Rafael (2014)
38. Munera, D.: Solving hard combinatorial optimization problems using cooperative parallel metaheuristics. Ph.D. thesis, University of Paris-1, France (2016)
39. Nethercote, N., Stuckey, P.J., Becket, R., Brand, S., Duck, G.J., Tack, G.: MiniZinc: towards a standard CP modelling language. In: Bessière, C. (ed.) CP 2007. LNCS, vol. 4741, pp. 529–543. Springer, Heidelberg (2007). https://doi.org/10.1007/978-3-540-74970-7_38

40. Okada, S., Ohzeki, M., Taguchi, S.: Efficient partition of integer optimization problems with one-hot encoding, September 2019
41. Oshiyama, H., Ohzeki, M.: Benchmark of quantum-inspired heuristic solvers for quadratic unconstrained binary optimization (2021)
42. Rodríguez-Heck, E.: Linear and quadratic reformulations of nonlinear optimization problems in binary variables. Ph.D. thesis, University of Liège, Belgium (2018)
43. Rosenberg, I.: Reduction of bivalent maximization to the quadratic case. Cahiers du Centre d'Etudes de Recherche Operationnelle **17**(71) (1975)
44. Russo, J., Erickson, K., Beard, J.: Costas array search technique that maximizes backtrack and symmetry exploitation. In: CISS, pp. 1–8 (2010)
45. Simonis, H.: Limits of propagation (costas array). ECLiPSe ELEarning slides. https://eclipseclp.org/ELearning/costas/handout.pdf
46. Stollenwerk, T., Lobe, E., Jung, M.: Flight gate assignment with a quantum annealer. In: Feld, S., Linnhoff-Popien, C. (eds.) QTOP 2019. LNCS, vol. 11413, pp. 99–110. Springer, Cham (2019). https://doi.org/10.1007/978-3-030-14082-3_9
47. Takemoto, T., et al.: A 144kb annealing system composed of 9×16kb annealing processor chips with scalable chip-to-chip connections for large-scale combinatorial optimization problems. In: 2021 IEEE International Solid-State Circuits Conference (ISSCC), pp. 64–66 (2021)
48. Tanaka, S., Tamura, R., Chakrabarti, B.K.: Quantum Spin Glasses, Annealing and Computation, 1st edn. Cambridge University Press, Cambridge (2017)
49. Tatsumura, K.: Large-scale combinatorial optimization in real-time systems by FPGA-based accelerators for simulated bifurcation. In: HEART21 - International Symposium on Highly Efficient Accelerators and Reconfigurable Technologies. ACM (2021)
50. Vellino, A.: Costas arrays. Technical report, Bell-Northern Research, Ottawa, Canada (1990)
51. Vulakh, D., Finkel, R.: Parallel m-dimensional relative ant colony optimization (mdraco) for the costas-array problem (2021). https://doi.org/10.21203/rs.3.rs-975983/v1, preprint
52. Wagner, Z.A., Garren, D.A., Pace, P.E.: SAR imagery via frequency shift keying costas coding. In: 2017 IEEE Radar Conference (RadarConf), pp. 1789–1792 (2017)
53. Yamaoka, M., Okuyama, T., Hayashi, M., Yoshimura, C., Takemoto, T.: CMOS annealing machine: an in-memory computing accelerator to process combinatorial optimization problems. In: IEEE Custom Integrated Circuits Conference, CICC 2019, Austin, TX, USA, 14–17 April 2019, pp. 1–8. IEEE (2019)

Penalty Weights in QUBO Formulations: Permutation Problems

Mayowa Ayodele[(✉)] [iD]

Fujitsu Research of Europe, Slough, UK
`mayowa.ayodele@fujitsu.com`

Abstract. Optimisation algorithms designed to work on quantum computers or other specialised hardware have been of research interest in recent years. Commercial solvers that use quantum or quantum-inspired methods, such as Fujitsu's Digital Annealer (DA) and D-wave's Quantum Annealer, can solve optimisation problems faster than algorithms implemented on general purpose computers. However, they can only optimise problems that are in binary and quadratic form. Quadratic Unconstrained Binary Optimisation (QUBO) is therefore a common formulation used by these solvers.

There are many combinatorial optimisation problems that are naturally represented as permutations e.g. travelling salesman problem. Encoding permutation problems using binary variables however presents some challenges. Many QUBO solvers are single flip solvers, it is therefore possible to generate solutions that cannot be decoded to a valid permutation. To create bias towards generating feasible solutions, we use penalty weights. The process of setting static penalty weights for various types of problems is not trivial. This is because values that are too small will lead to infeasible solutions being returned by the solver while values that are too large may lead to slower convergence. In this study, we explore some methods of setting penalty weights within the context of QUBO formulations. We propose new static methods of calculating penalty weights which lead to more promising results than existing methods.

Keywords: Quantum-Inspired Optimisation · Digital Annealer · Permutation · Penalty Weights · Constraint Handling · Quadratic Unconstrained Binary Optimisation · Ising Model · Binary Quadratic Problem

1 Background

Permutation problems are well studied combinatorial optimisation problems in nature inspired computing. They have many real-world applications especially in planning and logistics. Some of the most frequently studied permutation problems in literature are the well-known Travelling Salesman Problem (TSP) and Quadratic Assignment Problem (QAP). Since these problems are NP-hard, heuristics and meta-heuristics have been proposed for solving them.

© The Author(s), under exclusive license to Springer Nature Switzerland AG 2022
L. Pérez Cáceres and S. Verel (Eds.): EvoCOP 2022, LNCS 13222, pp. 159–174, 2022.
https://doi.org/10.1007/978-3-031-04148-8_11

Several classes of algorithms have been applied to problems naturally represented as permutations e.g. Estimation of Distribution Algorithm [2], Iterated Local Search and Differential Evolution Algorithm [21]. Quantum-inspired algorithms such as the Digital Annealer (DA) has also been shown to present more promising performance on the QAP when compared to CPLEX and QBSolve [16]. The DA was particularly shown to be up to three or four orders of magnitude faster than CPLEX on QAP instances and maximum cut instances.

Quantum and quantum-inspired methods have been of research interest in recent years. This is because they are able to exploit the use of specialised hardware to solve optimisation problems much quicker than classical algorithms implemented on general purpose machines [1]. It is common to formulate combinatorial optimisation problems such as permutation problems in quadratic and binary form such that algorithms that use specialised hardware including (but not limited to) quantum devices can be used to solve them. In recent years, Quadratic Unconstrained Binary Optimisation (QUBO) has become a unifying model for representing many combinatorial optimisation problems [24]. QUBO (or the equivalent Ising) formulations of common combinatorial optimisation problems are presented in [15]. As the name depicts, QUBO problems are unconstrained, quadratic and of binary form. Since the representation only supports binary information, the natural representation of permutation problems can therefore not be used. While some classical optimisation algorithms such as genetic algorithms use permutation representation [12] to solve QAP and/or TSP, other algorithms require alternative representations e.g. random keys [3], factoradics [18], binary [4], matrix representation [14] or two-way one-hot [15]. The two-way one-hot representation [8,15,16], also known as permutation matrix [5], is often used in QUBO formulations of permutation problems and is used in this study. In this representation, a substring of bits is used to represent each entity (e.g. a location in TSP). In each substring, only one bit can be turned on and in the entire solution, the bit turned on in each substring must be unique. This representation ensures that the mutual exclusivity constraint of permutation problems is respected. QUBO solvers such as Path Relinking method used in [24] and the first generation DA in [1] are single flip solvers and are therefore not able to preserve two-way one-hot validity. To ensure that the problem to be solved is in 'unconstrained' form, penalty weights are applied to combine the cost function (unconstrained objective function) with the constraint function. Solutions are penalised by the magnitude of violation of the constraint(s).

Setting penalty weights is not a trivial task. Values that are too large make the search too difficult for the solver as the penalty terms overwhelm the original objective function [24]. Penalty weights that are too small are highly undesirable as infeasible solutions will displace feasible solutions in the search, causing the solver to return infeasible solution(s). Penalty weights are however not unique and there are often a range of values that work well [9]. The primary objective of setting the penalty weights is often to ensure that the optimal solution of the QUBO is the optimal solution of the original constrained problem. However, it is also important that these values are not too large.

In literature, there are many approaches of setting penalty weights for QUBOs. The value can be set by domain experts [9]. A common approach is to derive the penalty weights empirically using methods that increase the weights until feasibility is achieved [20]. This approach is however computationally intensive as a full run of the algorithm and analysis of results is required each time until the right value is reached.

Another common approach is to set the penalty weight to a value greater than the largest possible objective value. Deriving the range (upper bound and lower bound) of the objective function is often problem specific. These values can also be too large. Although there are some general methods of determining the bounds of a QUBO, it is however often the case that methods with less computational complexity lead to values that are too large while methods that can provide better bounds are often computationally expensive [6]. Moreover, while better bounds can lead to smaller but valid penalty weights (i.e. guarantees feasibility of the optimal), the penalty values are still often larger than desired. In [22], the performance of the DA was analysed using different penalty weights that are fractions of the range of the objective function. The range was derived using problem specific information, the authors found much smaller values can lead to better performance of the DA.

Furthermore, the maximum coefficient of the QUBO has been used as penalty weights when solving problems like the TSP. The idea behind this is that, if a TSP solution is penalised by the maximum distance between any two cities, feasibility of the optimal solution can always be achieved. In [23], the authors used a multi-trial approach. The values used were within a range defined using the minimum and the maximum distance between any two locations. Similarly, in [10], fractions of the maximum distance were used to derive penalty weights for the TSP. This is an example of a scenario where values much lower than the full range of the objective function can be valid. This approach may however not generalise to other problems [24].

In [24], a pre-processing method that can be used to generate penalties for equality constraints within the context of single flip QUBO solvers was presented. The authors measure the maximum change in objective function that can be obtained as a result of any single flip in a solution. This method, which is referred to as VLM in this study, is explained in more details in Sect. 3.

Examples of QUBO (or Ising) solvers that use specialised hardware are D-wave's Quantum Annealer [13] and Fujitsu's DA [1]. The Quantum Annealer uses Quantum Processing Units (QPU) to achieve its speed up while the DA uses application-specific CMOS hardware. In this study, we analyse the effect of different methods of generating penalty weights for permutation problems (TSP and QAP). Initial experiments are based on a CPU implementation of the DA Algorithm presented in [1]. Further analysis of the effect of penalty weights are done using the third generation DA (DA3) as the QUBO solver [11].

The rest of this paper is structured as follows. Section 2 presents the permutation formulations and QUBO formulations of the TSP and QAP used in this study. Section 3 presents the methods of generating penalty weights. Section 4 presents a description of the DA Algorithm. Experimental Settings are presented in Sect. 5. Analyses of results and conclusion are presented in Sects. 6 and 7 respectively.

2   Permutation Problems

A valid permutations is described as $\sigma = \{\sigma_1, \ldots, \sigma_n\}$, where $\sigma_i \neq \sigma_j \ \forall \ i \neq j$. In general, QUBO problems can be defined as follows:

$$E(x) = x^T Q x + k , \tag{1}$$

where Q and k are $m \times m$ QUBO matrix and constant term, the solution $x = (x_1, \ldots, x_m)$ is an m-dimensional vector, and $E(x)$ is the energy (or fitness) of x. To formulate permutation problems as QUBO, it is important for the problem to be formulated as zeros and ones as these are the only values supported by QUBO solvers. Some of the well-known approaches of transforming integer values to zeros and ones are binary, gray code and one-hot. Within the context of QUBOs, the two-way one-hot (permutation matrix) encoding is often used to represent permutations and will be used in this study.

The energy (fitness function) of the problem is shown in Eq. (2) where $c(x)$ and $g(x)$ are respectively cost and constraint functions while α is the penalty weight. The generic cost and constraint functions of permutation problems are presented in Eqs. (3) and (4) [10].

$$\text{minimise } E(x) = c(x) + \alpha \times g(x) \tag{2}$$

$$c(x) = \sum_{i=1}^{n}\sum_{j=1}^{n}\sum_{k=1}^{n}\sum_{l=1}^{n} x_{i,k} Q_{i,k,j,l} x_{j,l} \tag{3}$$

$$g(x) = \sum_{i=1}^{n}\left(1 - \sum_{k=1}^{n} x_{i,k}\right)^2 + \sum_{k=1}^{n}\left(1 - \sum_{i=1}^{n} x_{i,k}\right)^2 . \tag{4}$$

Note that the constraint function is designed to ensure that the solutions can be converted to valid permutations i.e. penalise solutions that do not satisfy,

$$\sum_{k=1}^{n} x_{i,k} = 1 \ \forall \ i \in \{1, \ldots, n\}, \quad \sum_{i=1}^{n} x_{i,k} = 1 \ \forall \ k \in \{1, \ldots, n\} . \tag{5}$$

In Eqs. (3)–(5), binary variable $x_{i,k}$ indicates whether an object i is assigned to position k or not. We however note that x is solved as a vector of size $m = n^2$ rather than a $n \times n$ matrix. Also, while $Q_{i,k,j,l}$ in Eq. (3) is the QUBO coefficient that captures the relationship between an object i being in position k and an object j being in position l. In the rest of this paper, QUBO matrices C and G representing the cost or constraint functions are presented as $m \times m$ matrices.

2.1   Quadratic Assignment Problem

The QAP can be described as the problem of assigning a set of n facilities to a set of n locations. For each pair of locations, a distance is specified. For each pair of facilities, a flow (or weight) is specified. The aim is to assign each facility to a unique location such that the sum of the products between flows and distances is minimised.

Formally, the QAP consists of two $n \times n$ input matrices $H = [h_{i,j}]$ and $D = [d_{k,l}]$, where $h_{i,j}$ is the flow between facilities i and j, and $d_{k,l}$ is the distance between locations k and l, the solution to the QAP is a permutation $\sigma = (\sigma_1, \dots, \sigma_n)$ where σ_i represents the location that facility i is assigned to. The objective function (total cost) is formally defined as follows

$$\text{minimise } f(\sigma) = \sum_{i=1}^{n} \sum_{j=1}^{n} h_{i,j} \times d_{\sigma_i, \sigma_j} \ . \tag{6}$$

We aim to solve Eq. (2) where the cost function $c(x)$ of the QUBO representing the QAP is presented in Eq. (7) and the constraint function $g(x)$ is the same as Eq. (4).

$$c(x) = \sum_{i=1}^{n} \sum_{j=1}^{n} \sum_{k=1}^{n} \sum_{l=1}^{n} h_{i,j} d_{k,l} x_{i,k} x_{j,l} \ . \tag{7}$$

Any solution x can be encoded with n^2 variables, when x is presented in vector format.

2.2   Travelling Salesman Problem

The TSP consists of n locations and a matrix d representing distances between any two locations. The aim of the TSP is to minimise the distance travelled while visiting each location exactly once and returning to the location of origin. Given that σ_i is used to denote the ith city and $d_{\sigma_{i-1}, \sigma_i}$ is the distance between σ_i and σ_{i-1}. The solution to the TSP is a permutation $\sigma = \{\sigma_1, \dots, \sigma_n\}$ where each σ_i $(i = 1, \dots, n)$ represents the ith location to visit. The TSP is formally defined as

$$\text{minimise } f(\sigma) = \sum_{i=2}^{n} d_{\sigma_{i-1}, \sigma_i} + d_{\sigma_n, \sigma_1} \ . \tag{8}$$

We aim to solve Eq. (2) where the cost function, $c(x)$ of the QUBO representing the TSP is presented in Eq. (9) and the constraint function $g(x)$ is the same as Eq. (4).

$$c(x) = \sum_{(l,i) \in E} d_{l,i} \sum_{k=1}^{n} x_{l,k} x_{i,k+1} \ . \tag{9}$$

The TSP instances used in this work are symmetric, we can therefore fix the first city, reducing the size of x to $(n-1)^2$ [15].

Note that QUBOs for these problems can be generated using packages such as PyQUBO [25]. QUBOs are generated in this study using method in [17].

3   Penalty Weights

The aim of this study is to derive methods of setting α in Eq. (2) such that the optimal solution to the penalised objective function is the optimal solution of the original constrained problem. We do this without problem specific knowledge but use information captured in the QUBO matrices representing the cost and constraint functions. This is shown in Eq. (10), $c(y)$ is used to denote the cost function of the optimal solution y. S is the solution space of infeasible solutions. Note that $g(x)$ produces a non-negative value, $g(x) = 0$ if the solutions are feasible but $g(x) > 0$ for infeasible solutions. The value of $g(x)$ increases according to the degree of constraint violation.

$$c(y) < c(x) + \alpha \times g(x) \ \forall \ x \in S. \tag{10}$$

Eq. (10) implies that a valid penalty weight α is one that satisfies Eq. (11)

$$\alpha > \max_{x \in S} \left(\frac{c(y) - c(x)}{g(x)} \right). \tag{11}$$

In the rest of this study, C and G are used to denote the QUBO matrices representing $c(x)$ and $g(x)$ respectively. Note that Q, which is the QUBO matrix optimised by the solver, can be derived by aggregating the matrices (i.e. $Q = C + \alpha \times G$), where $\alpha \geq 1$. The methods of generating penalty weights used in this study are described as follow:

UB: A common method of setting penalty weights is based on the Upper Bound (UB) of the objective function. The UB of C used in this study is presented in Eq. (12). This is a valid upper bound for problems with all positive QUBO coefficients. We note that a solution consisting of all 1s is an infeasible solution but it gives an estimate of how large the objective function could be.

$$UB = z^T C z, \ z_i = 1 \ \forall \ i \ \in [1, n]. \tag{12}$$

MQC: The Maximum QUBO Coefficient (MQC) which also corresponds to the maximum distance between any two cities in the TSP has been used as penalty weights in previous study [15]. MQC is defined in Eq. (13).

$$MQC = \max_{i=1}^{n} \max_{j=1}^{n} C_{i,j}. \tag{13}$$

VLM: This is the method proposed by Verma and Lewis [24]. For a 1-flip solver like the DA, any variable x_i can be flipped from 0 to 1 and vice versa at each iteration of the algorithm. VLM focuses on deriving a good estimate for the numerator of Eq. (11), i.e. $(c(y) - c(x))$. The method estimates the amount of gain/loss in objective function that can be achieved by either turning a bit on or off. They do not consider the denominator (i.e. $g(x)$) which the authors recognise will be hard to estimate without complete enumeration. Since $g(x) > 0$ in infeasible solutions, VLM (Eq. (14)) will always be valid.

$$W^c = \left\{ -C_{i,i} - \sum_{\substack{j=1 \\ j \neq i}}^{n} \min\{C_{i,j}, 0\}, \ C_{i,i} + \sum_{\substack{j=1 \\ j \neq i}}^{n} \max\{C_{i,j}, 0\} \ \forall \ i \in [1, n] \right\} \quad (14)$$

$$\alpha = \text{VLM} = \max_{i=1}^{n} W_i^c . \quad (15)$$

MOMC: We propose an amendment to VLM [24]. We refer to the proposed method as the Maximum change in Objective function divided by Minimum Constraint function of infeasible solutions (MOMC). We note that $g(x)$ is not considered in Eq. (14). VLM can be reduced such that α is still valid, if we know the minimum constraint function $(g(x))$ of any infeasible solution. This can be computed from G by estimating the minimum change in constraint function that is greater than 0 as shown in Eq. (16).

$$W^g = \left\{ -G_{i,i} - \sum_{\substack{j=1 \\ j \neq i}}^{n} \min\{G_{i,j}, 0\}, \ G_{i,i} + \sum_{\substack{j=1 \\ j \neq i}}^{n} \max\{G_{i,j}, 0\} \ \forall \ i \in [1, n] \right\} \quad (16)$$

$$\gamma = \min_{\substack{i=1 \\ W_i^g > 0}}^{n} W_i^g . \quad (17)$$

For permutation problems represented as two-way one-hot, $g(x)$ of any solution that is a flip away from a feasible solution is 2 (i.e. $\gamma = 2$). Method of generation α using the proposed MOMC is presented in Eq. 18 where W^c and γ are derived as shown in Eqs. (14) and (17)

$$\alpha = \text{MOMC} = \max\left(1, \frac{\text{VLM}}{\gamma}\right) = \max\left(1, \frac{\max_{i=1}^{n} W_i^c}{2}\right) . \quad (18)$$

MOC: We propose another amendment to the VLM method. The method presented here is derived by selecting the Maximum value derived from dividing each change in Objective function with the corresponding change in Constraint function (MOC). MOC captures possible equivalent increase in constraint function as a result of a change in objective function which could be achieved by flipping any bit from 0 to 1 or vice versa. Method of generation α using the proposed MOC is presented in Eq. (19) where W^c and W^g are derived as shown in Eqs. (14) and (16)

$$\alpha = \text{MOC} = \max\left(1, \max_{\substack{i=1 \\ W_i^g > 0}}^{n} abs\left(\frac{W_i^c}{W_i^g}\right)\right) . \quad (19)$$

4 Digital Annealer

The DA is a technology designed to solve large scale combinatorial optimisation problems in much shorter time than most classical algorithms.

Algorithm 1. The DA (1st generation) Algorithm

1: initial_state ← an arbitrary state
2: **for each** run **do**
3:    initialise to initial_state
4:    E_{offset} ← 0
5:    **for each** iteration **do**
6:       update the temperature if due for temperature update
7:       **for each** variable j, in parallel **do**
8:          propose a flip using $\Delta E_j - E_{\text{offset}}$
9:          if acceptance criteria (P_j) is satisfied, record
10:      **end for**
11:      **if** at least one flip is recorded as meeting the acceptance criteria **then**
12:         chose one flip at random from recorded flips
13:         update the state and effective fields, in parallel
14:         E_{offset} ← 0
15:      **else**
16:         $E_{\text{offset}} = E_{\text{offset}} +$ offset_increase_rate
17:      **end if**
18:   **end for**
19: **end for**

The first generation DA is a single flip solver with similar properties as the Simulated Annealing (SA). It is however designed to be more effective than the classical SA algorithm [1]. The standard algorithm of the first generation DA is presented in Algorithm 1. The DA algorithm exploits the use of specialised hardware such that all neighbouring solutions are explored in parallel and in constant time regardless of the number of neighbours. This approach significantly improves acceptance probabilities of the regular SA algorithm. The DA does not completely evaluate each solution but computes the energy difference resulting from flipping any single bit of the parent solution. Also, the DA uses an escape mechanism to avoid being trapped in local optimal. As shown in the algorithm, E_{offset} is used to relax acceptance criteria. The E_{offset} is incremented by a parameter (offset_increase_rate) each time no neighbour which satisfies the acceptance criteria is found. The acceptance criteria used in this study is $P_j = \exp(\min(0, -(\Delta E_j - E_{\text{offset}})/\delta))$ where P_j is the probability of accepting the jth flip. Note that ΔE_j represents the difference in energy as a result of flipping the jth bit and δ is the current temperature.

More extensive details of the algorithm can be seen in [1]. The first and second generation DAs were released in May 2018 and December 2018. Both versions were designed to solve optimisation problems that have been formulated as QUBOs. The most recent generation of the DA is the DA3 which is able to find optimal or sub optimal solutions to Binary Quadratic Problems (BQP) of up to 100,000 bits [11]. BQPs include QUBO but also other binary and quadratic formulations that may have constraints. Note that the algorithm that supports DA3 has been updated to perform better than the current algorithm presented. For simplicity, we use the CPU implementation of Algorithm 1 in this study to evaluate the quality of solution derived when using different penalty weights. We however also presented some results derived using DA3.

5   Experimental Setup

Problem sets, measure of performance and parameter settings used in this study are presented in this section.

5.1   Datasets

In order to compare the behaviour of the DA with different penalty weights, we used common TSP and QAP instances from TSPLIB [19] and QAPLIB [7]. The instances used in this study and the corresponding solution sizes m are presented in Table 1.

Table 1. QAP and TSP instances and their solution sizes

QAP Instances	m	QAP Instances	m	TSP Instances	m	TSP Instances	m
had12	144	rou12	144	bays29	784	fri26	625
had14	196	rou15	225	bayg29	784	gr17	256
had16	256	rou20	400	berlin52	2601	gr21	400
had18	324	tai40a	1600	brazil58	3249	gr24	529
had20	400	tai40b	1600	dantzig42	1681	st70	4761

QUBO matrices (in upper triangular format) representing the cost and constraint functions of these QAP and TSP instances used are made available[1].

5.2   Performance Measure

We compare the performance of the DA using different methods of generating penalty weights. The performance measure used is the Average Relative Percentage Deviation (ARPD) defined in Eq. (20)

$$ARPD = \frac{1}{r} \sum_{i=1}^{r} \left(\frac{DA(\alpha)_i - \text{Optimal}}{\text{Optimal}} \right) \times 100 \,. \tag{20}$$

Note that $DA(\alpha)_i$ is the best energy returned by the DA for the ith run using penalty weight set to α and r is the number of runs. We set $r = 20$ and the optimal value are obtained from QAPLIB and TSPLIB.

---

[1] https://github.com/mayoayodelefujitsu/QUBOs.

5.3 Parameter Settings: DA Algorithm

The parameter settings used in the DA and DA3 are shown in Table 2. For the DA, the temperature is set to decrease from 'Initial Temperature' to 'Final Temperature' by a fraction 'Temperature Decay'. Note $\delta_i = \max(\delta_f, \delta_{i-1} * (1 - \rho))$ where δ_i denotes the temperature at iteration i. In the DA3, temperature and offset increment related parameters are automatically set. Therefore, no manual setting is required for the DA3, these parameters are thus shown as 'NA' for DA3 in Table 2. The stopping criteria used in the DA (CPU implementation) is 'number of iterations' but 'time (in seconds)' is used in DA3. This is because the two stopping criteria allowed in the DA3 are time and target energy (fitness). The DA is executed for m^2 iterations while the time limit for DA3 is set to the ceiling of 3% of m in seconds (m is presented in Table 1). Each experiment is executed independently 20 times. The parameters are chosen based on preliminary experiments. Note that $\beta = VLM$ (Eq. (14)).

Table 2. Parameter Settings

Parameter	DA	DA3
Initial Temperature δ_0	0.1β, β, 10β	NA
Final Temperature δ_f	1	NA
Stopping Criteria	m^2	$\lceil 0.03m \rceil$ sec
offset_increase_rate	$\delta_0 \div m^2$	NA
Number of Runs r	20	20
Temperature Decay ρ	0.001	NA

6 Results

In this section, different methods of generating penalty weights are compared using the parameter settings presented in Sect. 5.3. Results using the CPU implementation of the first generation DA algorithm are presented. Further results relating to the third generation DA are also presented.

6.1 Penalty Weights for TSP and QAP instances

Table 3 shows the penalty weights derived for different TSP and QAP instances using the methods of generating penalty weights defined in Sect. 3. The smallest and valid penalty weights are highlighted in bold. Validity of the penalty weights are assessed in Sect. 6.2. It should be noted that the methods were applied to QUBO matrices in upper triangular format.

Table 3. Penalty weights for QAP and TSP instances derived using different methods

Problem Category	Instance Name	Penalty Weights				
		UB	MQC	VLM	MOMC	MOC
QAP	had12	249,240	126	5,460	2,730	**488**
QAP	had14	573,484	162	8,968	4,484	**533**
QAP	had16	1,014,488	162	12,580	6,290	**545**
QAP	had18	1,832,940	200	16,102	8,051	**1,513**
QAP	had20	2,950,640	220	20,928	10,464	**1,335**
QAP	rou12	40,734,756	19,602	874,944	437,472	**34,531**
QAP	rou15	98,340,328	19,602	1,498,176	749,088	**79,715**
QAP	rou20	346,044,384	19,602	2,569,174	1,284,587	**123,342**
QAP	tai40a	5,904,547,332	19,602	10,418,804	5,209,402	**176,904**
QAP	tai40b	1,767,388,016,312	32,656,592	4,524,144,275	2,262,072,138	**56,133,309**
TSP	bayg29	3,381,534	**386**	6,279	3,140	2,404
TSP	bays29	4,259,764	**509**	8,593	4,297	3,003
TSP	berlin52	74,165,126	**1,716**	55,515	27,758	27,148
TSP	brazil58	379,655,572	**8,700**	288,552	144,276	55,557
TSP	dantzig42	4,814,472	**192**	5,029	2,515	1,915
TSP	fri26	1,455,150	**280**	4,833	2,417	1,616
TSP	gr17	1,005,188	**745**	7,981	3,991	3,074
TSP	gr21	2,666,064	**865**	11,160	5,580	2,853
TSP	gr24	1,609,942	**389**	5,185	2,593	1,888
TSP	st70	16,647,424	**129**	5,055	2,528	2,079

6.2 CPU Implementation of the DA Algorithm: Comparing Methods of Generating Penalty Weights

In this section, results derived using CPU implementation of the first generation DA algorithm are presented. Table 4 shows the number of feasible solutions returned by the DA within the stopping criteria when different methods of generating penalty weights are used.

Tables 5, 6 and 7 respectively present the ARPD derived by the DA at initial temperature set to 0.1β, β and 10β using different methods of generating penalty weights. The ARPD for MQC is not presented for QAP instances in any of the tables because no feasible solution was found.

UB, MQC and MOC present their best ARPD averaged across TSP instances when the temperature is set to 0.1β while MOMC and VLM present their best ARPD on TSP instances when temperature is set to β. UB presents its best performance on QAP instances when temperature is set to 0.1β, and MOC, MOMC and VLM present the best ARPD averaged across all QAP instances when the temperature is set to 10β.

In Table 5, the DA presents the best average ARPD on QAP instances when the method of setting penalty weight is set to MOC. The DA however presents the best average ARPD on TSP instances when MQC is used. These results are expected since the MOC and MQC respective present the smallest yet valid penalty weights for QAP and TSP instances. ARPD for QAP instance *rou12*, is shown in italics when the method of setting penalty weight is set to MOC because the ARPD is only computed using the 13 feasible solutions found while

Table 4. Number of DA runs that returned feasible solutions out of 20 runs using different methods of generating weights for QAP instances (left) and TSP instances (right). The same number of feasible solutions was obtained for QAP and TSP instances with different values of initial temperature ($\delta_0 = 0.1\beta/\beta/10\beta$) apart from 'rou12' when MOC is used, the respective values derived using each temperature is therefore presented.

Instance	Number of feasible runs					Instance	Number of feasible runs				
Name	UB	MQC	VLM	MOMC	MOC	Name	UB	MQC	VLM	MOMC	MOC
had12	20	0	20	20	20	bayg29	20	20	20	20	20
had14	20	0	20	20	20	bays29	20	20	20	20	20
had16	20	0	20	20	20	berlin52	20	20	20	20	20
had18	20	0	20	20	20	brazil58	20	20	20	20	20
had20	20	0	20	20	20	dantzig42	20	20	20	20	20
rou12	20	0	20	20	13/14/14	fri26	20	20	20	20	20
rou15	20	0	20	20	20	gr17	20	20	20	20	20
rou20	20	0	20	20	20	gr21	20	20	20	20	20
tai40a	20	0	20	20	20	gr24	20	20	20	20	20
tai40b	20	0	20	20	20	st70	20	20	20	20	20

Table 5. ARPD from Optimal on QAP (left) and TSP (right) instances where initial temperature $\delta_0 = 0.1 \times \beta$

Instance	Optimal	ARPD				Instance	Optimal	ARPD				
Name		UB	VLM	MOMC	MOC	Name		UB	MQC	VLM	MOMC	MOC
had12	1,652	14.15	12.98	11.98	**6.40**	bayg29	1,610	189.59	**52.94**	180.69	125.19	114.98
had14	2,724	16.20	14.85	13.86	**6.28**	bays29	2,020	190.45	**55.52**	188.23	130.47	116.61
had16	3,720	12.23	13.63	10.76	**5.50**	berlin52	7,542	295.70	**100.40**	289.46	212.58	209.45
had18	5,358	11.97	11.24	9.25	**6.35**	brazil58	25,395	389.04	**138.94**	390.04	276.79	260.21
had20	6,922	12.46	12.15	8.99	**6.25**	dantzig42	699	340.26	**95.05**	335.62	225.04	224.17
rou12	235,528	29.12	29.15	27.98	*10.37*	fri26	937	177.06	**57.32**	177.34	120.84	107.51
rou15	354,210	30.75	33.34	28.21	**16.28**	gr17	2,085	112.06	**29.67**	107.56	84.75	66.97
rou20	725,522	24.04	25.96	20.42	**14.35**	gr21	2,707	170.91	**44.82**	166.01	123.01	93.32
tai40a	3,139,370	20.44	20.73	16.08	**13.00**	gr24	1,272	166.45	**52.37**	160.64	114.54	102.29
tai40b	637,250,948	77.76	76.51	52.92	**11.73**	st70	675	444.14	**124.52**	419.62	330.44	325.83
Avg		24.91	25.05	20.05	**9.65**	Avg		247.57	75.16	241.52	174.37	162.13

others are generated using 20 feasible solutions. The DA presents the worst average ARPD on QAP (or TSP) instances when the method of setting penalty weights is set to VLM (or UB).

In Tables 6 and 7, similar to the results produced in Table 5, the DA presents the best average ARPD on QAP instances when the methods of setting penalty weight is set to MOC and the best average ARPD on TSP instances when set to MQC. ARPD for QAP instance *rou12*, is shown in italics when the method

Table 6. ARPD from Optimal on QAP (left) and TSP (right) instances where initial temperature $\delta_0 = \beta$

Instance Name	Optimal	ARPD				Instance Name	Optimal	ARPD				
		UB	VLM	MOMC	MOC			UB	MQC	VLM	MOMC	MOC
had12	1,652	15.25	7.65	8.33	**6.54**	bayg29	1,610	194.42	**54.69**	127.05	120.22	117.59
had14	2,724	15.26	9.37	9.76	**6.43**	bays29	2,020	196.69	**57.22**	124.98	122.92	120.10
had16	3,720	13.27	8.13	8.75	**5.41**	berlin52	7,542	298.15	**102.47**	217.57	214.73	214.16
had18	5,358	11.40	7.08	7.04	**6.55**	brazil58	25,395	380.88	**137.75**	278.99	277.66	261.90
had20	6,922	12.86	7.38	7.66	**6.74**	dantzig42	699	350.45	**100.25**	238.07	232.41	222.18
rou12	235,528	32.12	20.34	20.75	*9.58*	fri26	937	185.14	**59.04**	116.29	114.82	109.82
rou15	354,210	31.33	22.00	21.37	**15.75**	gr17	2,085	128.08	**31.41**	70.44	62.56	60.52
rou20	725,522	24.49	17.77	17.91	**13.69**	gr21	2,707	190.91	**52.99**	114.31	105.63	98.48
tai40a	3,139,370	20.43	16.10	16.13	**12.89**	gr24	1,272	178.25	**52.85**	114.77	104.18	98.75
tai40b	637,250,948	78.61	51.81	51.25	**11.49**	st70	675	435.61	**129.66**	335.66	331.78	329.34
Avg		25.50	16.76	16.89	**9.51**	Avg		253.86	**77.83**	173.81	168.69	163.28

Table 7. ARPD from Optimal on QAP (left) and TSP (right) instances where initial temperature $\delta_0 = 10 \times \beta$

Instance Name	Optimal	ARPD				Instance Name	Optimal	ARPD				
		UB	VLM	MOMC	MOC			UB	MQC	VLM	MOMC	MOC
had12	1,652	11.26	7.99	8.51	**6.22**	bayg29	1,610	193.29	**57.27**	127.64	122.24	122.83
had14	2,724	15.56	9.13	9.48	**6.11**	bays29	2,020	196.84	**57.57**	132.84	124.92	111.88
had16	3,720	14.02	8.19	8.19	**5.12**	berlin52	7,542	300.65	**103.48**	218.34	215.59	214.92
had18	5,358	11.80	7.07	7.31	**6.03**	brazil58	25,395	375.04	**139.30**	285.40	273.91	265.80
had20	6,922	12.57	7.32	7.33	**6.43**	dantzig42	699	333.03	**100.39**	238.24	227.12	230.27
rou12	235,528	28.30	18.94	16.50	*10.02*	fri26	937	180.85	**62.51**	124.09	114.74	11.27
rou15	354,210	33.98	21.02	20.16	**14.57**	gr17	2,085	122.95	**30.19**	70.65	64.84	60.17
rou20	725,522	25.19	17.80	17.36	**13.05**	gr21	2,707	178.13	**52.73**	115.05	107.30	99.54
tai40a	3,139,370	20.96	16.00	15.97	**12.54**	gr24	1,272	179.79	**56.82**	116.19	106.69	103.71
tai40b	637,250,948	79.65	50.85	49.94	**12.10**	st70	675	452.83	**126.34**	337.61	334.76	325.69
Avg		25.33	16.43	16.07	**9.22**	Avg		251.34	**78.66**	176.61	169.21	164.56

of setting penalty weight is set to MOC because the ARPD is only computed using the 14 feasible solutions found while others are generated using 20 feasible solutions. The DA presents the worst average ARPD on QAP and TSP instances when the method of setting penalty weights is set to UB.

In general, the results show that the methods which produced the smallest valid penalty weights (MQC for TSP and MOC for QAP) consistently produced the best ARPD. Conversely, the results also show that the method that produced the largest penalty weights on TSP and QAP instances (UB) often presents the worst ARPD. For permutation problems represented as two-way one-hot, all neighbours of a feasible solution are infeasible solutions. It is therefore important for penalty weights to be small enough to encourage the algorithm to explore infeasible solutions in order to find better feasible solutions.

6.3   DA3: Comparing Methods of Generating Penalty Weights

In Sect. 6.2, we show how different methods of generating penalty weights can affect the performance of the 1st generation DA (CPU implementation). In this section, we present results using DA3 (i.e. version of the third generation DA which benefits from hardware speed-up). It should be noted that the DA3 has more capabilities and can be executed in many modes. A major improvement presented by the DA3 is the ability to handle linear inequality and one-hot constraints.

Table 8. ARPD derived using DA3 with different methods of generating penalty weights on QAP (left) and TSP (right) instances

Instance Name	Optimal	ARPD				Instance Name	Optimal	ARPD				
		UB	VLM	MOMC	MOC			UB	MQC	VLM	MOMC	MOC
had12	1,652	0.00	0.00	0.00	0.00	bayg29	1,610	0.00	0.00	0.00	0.00	0.00
had14	2,724	0.00	0.00	0.00	0.00	bays29	2,020	0.00	0.00	0.00	0.00	0.00
had16	3,720	0.00	0.00	0.00	0.00	berlin52	7,542	**1.86**	2.96	4.16	6.05	2.20
had18	5,358	0.00	0.00	0.00	0.00	brazil58	25,395	1.53	3.33	1.53	1.58	**1.43**
had20	6,922	0.00	0.00	0.00	0.00	dantzig42	699	0.00	0.00	0.00	0.00	0.00
rou12	235,528	0.00	0.00	0.00	0.00	fri26	937	0.00	0.00	0.00	0.00	0.00
rou15	354,210	0.00	0.00	0.00	0.00	gr17	2,085	0.00	0.00	0.00	0.00	0.00
rou20	725,522	0.00	0.00	0.00	0.00	gr21	2,707	0.00	0.00	0.00	0.00	0.00
tai40a	3,139,370	0.07	0.07	0.07	0.07	gr24	1,272	0.00	0.00	0.00	0.00	0.00
tai40b	637,250,948	0.00	0.00	0.00	0.00	st70	675	2.67	**1.48**	2.41	2.37	2.39
Average		0.01	0.01	0.01	0.01	Average		0.61	0.78	0.81	1.00	0.60

We present the ARPD achieved within 0.03 m seconds (where m represents the size of the solution) of executing DA3 with different methods of generating penalty weights in Table 8. DA3 achieves 100% feasibility on TSP instances with any of the penalty methods, it also achieves 100% feasibility on QAP when UB, VLM, MOMC and MOC methods are used. Furthermore, the standard deviation across 20 runs of the DA3 is often 0. For all QAP instances as well as 7 out of 10 TSP instances, DA3 obtains the same ARPD regardless of the penalty weights. DA3 presents varying ARPD on the largest TSP instances when different methods of generating penalty weights are used. Best ARPD was obtained for berlin52, brazil58 or st70 when UB, MOC or MQC is used respectively. There is therefore no clear evidence of one method being the best. We can therefore not make the same conclusions as the previous section, where there was clear evidence of smaller and valid penalty weights leading to better solution quality. Similarities in performance of DA3 regardless of penalty weights used may be because of the algorithmic changes made since 1st generation DA. DA3 is designed to solve problems formulated as two-way one-hot more efficiently [11]. It is also able to automatically find the best parameter settings for any BQP. The algorithm that supports DA3 is however not publicly available, it is therefore difficult to be precise about the reason for the difference in performance when compared to the first generation DA.

7   Conclusion and Further Work

Permutation problems like TSP and QAP can be formulated as QUBO such that algorithms that use specialised hardware e.g. Quantum Annealer or DA can solve them. Transforming these problems into QUBO form requires the setting of penalty weights. In this study, we examined different methods of generating penalty weights within the context of using the DA algorithm for solving permutation problems. The permutation problems used are TSP and QAP. We present improvements to existing methods of generating penalty weights leading to better performance of the DA. Although the DA algorithm, which shares similar properties with SA, was influenced by the magnitude of penalty weights, we could not reach the same conclusions with DA3. It was impossible to do deeper analysis because the DA3 algorithm is not publicly available. Further research into how various algorithms behave with different mechanisms of generating penalty weights is therefore necessary.

References

1. Aramon, M., Rosenberg, G., Valiante, E., Miyazawa, T., Tamura, H., Katzgraber, H.G.: Physics-inspired optimization for quadratic unconstrained problems using a digital annealer. Front. Phys. **7**, 48 (2019)
2. Arza, E., Pérez, A., Irurozki, E., Ceberio, J.: Kernels of mallows models under the hamming distance for solving the quadratic assignment problem. Swarm Evol. Comput. **59**, 100740 (2020)
3. Ayodele, M., McCall, J., Regnier-Coudert, O.: RK-EDA: a novel random key based estimation of distribution algorithm. In: Handl, J., Hart, E., Lewis, P.R., López-Ibáñez, M., Ochoa, G., Paechter, B. (eds.) PPSN 2016. LNCS, vol. 9921, pp. 849–858. Springer, Cham (2016). https://doi.org/10.1007/978-3-319-45823-6_79
4. Baluja, S.: Population-based incremental learning. a method for integrating genetic search based function optimization and competitive learning. Technical report, Department of Computer Science, Carnegie-Mellon University, Pittsburgh (1994)
5. Birdal, T., Golyanik, V., Theobalt, C., Guibas, L.J.: Quantum permutation synchronization. In: Proceedings of the IEEE/CVF Conference on Computer Vision and Pattern Recognition (CVPR), pp. 13122–13133, June 2021
6. Boros, E., Hammer, P.L., Sun, R., Tavares, G.: A max-flow approach to improved lower bounds for quadratic unconstrained binary optimization (QUBO). Discret. Optim. **5**(2), 501–529 (2008)
7. Burkard, R.E., Karisch, S.E., Rendl, F.: QAPLIB-a quadratic assignment problem library. J. Global Optim. **10**(4), 391–403 (1997)
8. Glover, F., Kochenberger, G., Du, Y.: A tutorial on formulating and using QUBO models. arXiv preprint arXiv:1811.11538 (2018)
9. Glover, F., Kochenberger, G., Du, Y.: Quantum bridge analytics i: a tutorial on formulating and using QUBO models. 4OR **17**(4), 335–371 (2019)

10. Goh, S.T., Gopalakrishnan, S., Bo, J., Lau, H.C.: A hybrid framework using a QUBO solver for permutation-based combinatorial optimization. arXiv preprint arXiv:2009.12767 (2020)

11. Hiroshi, N., Junpei, K., Noboru, Y., Toshiyuki, M.: Third generation digital annealer technology (2021). https://www.fujitsu.com/global/documents/about/research/techintro/3rd-g-da_en.pdf

12. Hussain, A., Muhammad, Y.S., Nauman Sajid, M., Hussain, I., Mohamd Shoukry, A., Gani, S.: Genetic algorithm for traveling salesman problem with modified cycle crossover operator. Comput. Intell. Neurosci. **2017** (2017)

13. Johnson, M.W., et al.: Quantum annealing with manufactured spins. Nature **473**(7346), 194–198 (2011)

14. Larranaga, P., Kuijpers, C.M.H., Murga, R.H., Inza, I., Dizdarevic, S.: Genetic algorithms for the travelling salesman problem: a review of representations and operators. Artif. Intell. Rev. **13**(2), 129–170 (1999)

15. Lucas, A.: Ising formulations of many np problems. Front. Phys. **2**, 5 (2014)

16. Matsubara, S., et al.: Digital annealer for high-speed solving of combinatorial optimization problems and its applications. In: 2020 25th Asia and South Pacific Design Automation Conference (ASP-DAC), pp. 667–672. IEEE (2020)

17. Moraglio, A., Georgescu, S.: Ising machine data input apparatus and method of inputting data into an ising machine, December 2020. https://worldwide.espacenet.com/patent/search?q=pn%3DEP3754564A1. Patent No. EP3754564A1, Filed 21st June 2019, Issued 9th August 2009

18. Regnier-Coudert, O., McCall, J.: Factoradic representation for permutation optimisation. In: Bartz-Beielstein, T., Branke, J., Filipič, B., Smith, J. (eds.) PPSN 2014. LNCS, vol. 8672, pp. 332–341. Springer, Cham (2014). https://doi.org/10.1007/978-3-319-10762-2_33

19. Reinelt, G.: TSPLIB-a traveling salesman problem library. ORSA J. Comput. **3**(4), 376–384 (1991)

20. Rosenberg, G., Haghnegahdar, P., Goddard, P., Carr, P., Wu, K., De Prado, M.L.: Solving the optimal trading trajectory problem using a quantum annealer. IEEE J. Sel. Top. Signal Process. **10**(6), 1053–1060 (2016)

21. Santucci, V., Baioletti, M., Milani, A.: Algebraic differential evolution algorithm for the permutation flowshop scheduling problem with total flowtime criterion. IEEE Trans. Evol. Comput. **20**(5), 682–694 (2015)

22. Şeker, O., Tanoumand, N., Bodur, M.: Digital annealer for quadratic unconstrained binary optimization: a comparative performance analysis. arXiv preprint arXiv:2012.12264 (2020)

23. Takehara, K., Oku, D., Matsuda, Y., Tanaka, S., Togawa, N.: A multiple coefficients trial method to solve combinatorial optimization problems for simulated-annealing-based ising machines. In: 2019 IEEE 9th International Conference on Consumer Electronics (ICCE-Berlin), pp. 64–69. IEEE (2019)

24. Verma, A., Lewis, M.: Penalty and partitioning techniques to improve performance of QUBO solvers. Discrete Optim. 100594 (2020). https://doi.org/10.1016/j.disopt.2020.100594. https://www.sciencedirect.com/science/article/pii/S1572528620300281. Issn: 1572-5286

25. Zaman, M., Tanahashi, K., Tanaka, S.: Pyqubo: python library for mapping combinatorial optimization problems to QUBO form. arXiv preprint arXiv:2103.01708 (2021)

PUBO$_i$: A Tunable Benchmark with Variable Importance

Sara Tari$^{(\boxtimes)}$, Sébastien Verel, and Mahmoud Omidvar

Univ. Littoral Côte d'Opale, UR 4491, LISIC, Laboratoire d'Informatique
Signal et Image de la Côte d'Opale, 62100 Calais, France
{sara.tari,verel,omidvar}@univ-littoral.fr

Abstract. In this work, we present the benchmark generator PUBO$_i$, Polynomial Unconstrained Binary Optimization, that combines subproblems to create instances of pseudo-boolean optimization problems. Any mono-objective pseudoboolean functions including existing classical optimization problems can be expressed with Walsh functions. The benchmark generator can tune main features of problems such as problem dimension, non-linearity degree, and neutrality. Additionally, to be able to create instances with properties similar to those of real-like combinatorial optimization problems, the goal of PUBO$_i$ is to introduce the notion of variable importance. Indeed, the importance of decision variables can be tuned using three benchmark parameters. In the version presented here, we consider four subproblems already used in Chook generator for benchmarking quantum computers and algorithms as a basis. We also present the impact of benchmark parameters using a fitness landscape analysis that empirically shows these parameters to significantly impact the variable importance.

Keywords: Benchmark · Fitness landscape · Walsh function · Variable importance

1   Introduction

There exist numerous evolutionary algorithms, or local search algorithms that efficiently tackle combinatorial optimization problems. One main practical, and theoretical difficulty facing new problem instances is the selection, or the design of efficient optimization algorithm according to the properties of the instance to optimize. Studying such algorithms often require to run them on diverse problem instances, which is usually done successfully in benchmarking studies [1]. This paper attempts to propose a benchmark of pseudo-Boolean functions with tunable relevant properties.

In conjunction of benchmarking efforts, a powerful approach aiming to improve the understanding of optimization algorithms, and thus determine which approach to consider, consists of using fitness landscapes [28]. In evolutionary computation, a fitness landscape represents the search space and fitness function of a given instance according to the links between solutions induced by the

L. Pérez Cáceres and S. Verel (Eds.): EvoCOP 2022, LNCS 13222, pp. 175–190, 2022.
https://doi.org/10.1007/978-3-031-04148-8_12

neighborhood relation of the algorithm under consideration. This representation corresponds to a graph from which several characteristics can be estimated by the application of indicators on a sample of solutions [14,15], allowing a visualization of landscapes and features that induce challenges for optimization methods. Hence, fitness landscapes is a powerful tool to better guide the design and use of neighborhood-based algorithms, even on large landscapes.

To design new optimization algorithms by "hand" or using machine learning techniques, to test existing approaches, or to better understand problem difficulty using fitness landscape analysis often requires a large and diverse set of problem instances with relevant properties. To this aim, there exist several generators or benchmarks of academic problems. While a small subset of such instances are based on real-world data, most of them are randomly generated, with characteristics that could greatly differ from real-world problems. One of the peculiarities of real-like instances is that, contrary to instances generated using random-based techniques, they are structured. Thus some variables of the problems play a more important role in terms of fitness contribution, and can be interdependent to several other variables, making them harder to study and understand [12]. Although variable importance of such problems must be studied, existing benchmark generators do not allow users to tune this parameter, which contributes to the lack of instances having important variables.

In this work, we propose and present $PUBO_i$, a benchmark in which variable importance is tunable, and thus impacts the instance structure. $PUBO_i$ is based upon Walsh functions [25], an orthogonal basis of pseudo-boolean functions for representing any pseudo-boolean function. These functions arouse a strong interest in several scientific communities. In quantum physics, one can use these functions to create benchmarks representing optimization problems that can be tackled using new quantum computers [16]. They are successfully used in black-box combinatorial optimization as surrogate functions for computational-costly fitness functions [24]. Walsh functions also allow a standardized reformulation of various academic optimization problems [7]. This basis of functions can be used for benchmarking purposes, by generating instances with various characteristics, including characteristics similar to those of real-like instances such as the variable importance. The contribution of this paper are (1) the proposition of a new benchmark generator with tunable variable importance, (2) a fitness landscape analysis of the instances set aiming to analyze the impact of benchmark parameters on the shape of fitness landscape, and on the related problem difficulty.

The paper is organized as follows. The next section presents related works on benchmarking and variable importance. Section 3 is devoted to the description of $PUBO_i$ as well as the instances considered in this study. In Sect. 4, we present the methodology and results of our experiments. The last section provides a discussion of this work and points out future work directions.

2   Related Works

In this section we propose a brief overview of previous work on variable importance, test suites for optimization problems and benchmark generators, with a particular focus on a generator proposed by physicists we used as a basis for PUBO$_i$.

2.1   Variable Importance, and Benchmarks of Optimization Problems

In the context of optimization problems, we assume a variable is important when its mutation strongly affects the fitness of a solution. When this characteristic is present on instances to optimize, taking it into account can help to optimize the problem more efficiently. There exist a few studies of optimization algorithms focusing on the variable importance of optimization problems, although this characteristic often exists on real data. For example, in [17,18] the authors study a machine learning-enhanced recombination that incorporates an intelligent variable selection method for multi-objective optimization and show that taking the variable importance into consideration can improve the optimization quality. Another example concerns the fitness landscape analysis of landscapes from the SIALAC benchmark, a benchmark for mobility problems [12]. In this benchmark based on a city mobility simulation, variables have different degrees of importance, and a bandit descent heuristic is proposed to take important variables into account.

An efficient way of studying optimization algorithms consists of analyzing their behavior on various problems and instances. Over the past decades, a huge effort has been devoted to the proposition of benchmarks of various optimization problems. Benchmarks allow the community to focus on the optimization of a given problem and to study more efficiently various optimization problems. Among classic benchmarks for academic combinatorial optimization problems, Taillard [20] proposes instances covering three basic scheduling problems : the permutation flow shop, the job shop and the open shop scheduling problems. Despite their simplicity compared to real-life scheduling instances, many of these instances are still considered challenging and are widely used to study the flow-shop scheduling problem, both in terms of optimization methods and fitness landscape analysis. Another widely-used testbed is the QAPLIB, that provides several instances of the Quadratic Assignment Problem (QAP) [2], as QAP is a non-trivial problem for which a lot of effort has been devoted. QAPLIB comprises randomly-generated instances as well as instances based on real-life data, such as the testing of self-testable sequential circuits or the flow of patients between different facilities in a hospital. DIMACS instances for the graph coloring problem and maximal clique problem are provided in [11]. The families of provided graphs include random graphs, flat graphs or latin square graphs. While this list of instances libraries is not exhaustive, such libraries focus on a single problem or a small set of the same family of problems, and rarely propose a sufficient number of instances differing in variable importance.

In the field of continuous optimization, testbeds are regularly proposed for the GECCO workshop on Black-Box Optimization Benchmarking (BBOB) [9]. These testbeds can be found on the widely-recognized benchmarking software COCO [10] for black-box optimization. In addition to a noisy suite, COCO has been recently extended to multiobjective [23] and mixed integer [22] problems.

Recently, a benchmarking platform *IOHprofiler* has been proposed for evaluating the performance of iterative optimization heuristics [5]. One of its components, *IOHexperimenter* aims to generate benchmark suites for the benchmarking of such optimization methods. IOHexperimenter covers 23 real-valued (continuous), as well as 25 academic pseudo-boolean problems [5].

The aforementioned platforms implement a large panel of problems and are widely-used. These problems are either continuous or present a uniform structure, in which the weight of variables is generally uniform. While such platforms are particularly useful to study heuristic optimization methods, these studies are conducted on structures that are more common on academic problems than on real-ones, and thus cannot efficiently focus on the peculiarities induced by important variables.

2.2   Tile Planting Instances: Chook Generator

Perera *et al.* [16] proposed *Chook* a python-based generator that generates instances for the Tile Planting (TP) problem [8]. The goal of this benchmark is to test new quantum computers and algorithms. Indeed, combinatorial problems that can be considered efficiently with quantum computers are expressed as spinglasses problems, and even more particularly as Quadratic Unconstrained Binary Optimization problems (QUBO). TP problems are a benchmark of pseudoboolean objective functions (binary string search space) with known global minima, the planting solutions. These functions are determined by a sum of subfunctions. In the most general form, the fitness function (Hamiltonian) is:

$$\mathcal{H}(s) = \sum_{j \in V} h_j s_j + \sum_{k=2}^{n} \sum_{(i_1,\dots,i_k) \in E} J_{i_1,\dots,i_k} s_{i_1} s_{i_2} \dots s_{i_k}$$

with $s = (s_1, s_2, \dots, s_n)$, $s_i \in \{-1, 1\}$ called spins, and where the hypergraph $G = (V, E)$ with vertices V and edges E describes the interaction between the problem spins (variables). The coefficients $h_j \in \mathbb{R}$ define the local external field (energy between the environment and each spin), and the coefficients $J \in \mathbb{R}$ the intensity of each interaction. Notice that this equation can be mapped into boolean variables $x_i \in \{0, 1\}$ using the classical transformation $x_i = \frac{1}{2}(1 - s_i)$. In most cases these problems can be expressed using only 2-local interactions between spins with a graph G:

$$\mathcal{H}(s) = \sum_{j \in V} h_j s_j + \sum_{(i,j) \in E} J_{ij} s_i s_j$$

In TP instances [8], the problem graph is decomposed into edge-disjoint and vertex-sharing subgraphs. For a decomposition of the graph $G = (V, E)$ into

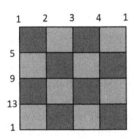

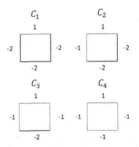

Fig. 1. Depiction of the graph for a 4 × 4 Tile Planting problem instance.

Fig. 2. Description of the four sub-function classes for the Tile Planting in Chook.

subgraphs $\{G_l = (V_l, E_l)\}$ such that no edges are shared among the subgraphs, the coefficients h_i are equal to 0, and each subgraph is associated to a quadratic energy function: $\mathcal{H}_l(s) = \sum_{(i,j) \in E_l} J_{ij} s_i s_j$

The energy function of the tile planting problem can then be expressed as the sum of energy functions of each sub-graph: $\mathcal{H} = \sum_l \mathcal{H}_l$

The sub-functions $\mathcal{H}_l$ are designed to share a common ground state, *i.e.* the lowest-energy state of a quantum-mechanical system, which corresponds to the ferromagnetic ground state $s = (+1, +1, \ldots, +1)$. Thus by the additive property of $\mathcal{H}$, the Tile Plating problems also have the same ferromagnetic ground state.

TP problems use a regular lattice structure that allows a decomposition of the graph G which contains a subset of the unit cells as subgraphs. In the square lattices version of the problem, the resulting unit-cells form a checkboard pattern (see Fig. 1) and the problem graph corresponds to a toric square matrix. A portfolio of four sub-function classes $\{C_1, C_2, C_3, C_4\}$ (see Fig. 2) are defined as follows: $C_1(s) = -2s_0 s_1 - 2s_1 s_2 - 2s_2 s_3 + s_3 s_0$; $C_2(s) = -2s_0 s_1 - 2s_1 s_2 - s_2 s_3 + s_3 s_0$; $C_3(s) = -2s_0 s_1 - s_1 s_2 - s_2 s_3 + s_3 s_0$; $C_4(s) = -s_0 s_1 - s_1 s_2 - s_2 s_3 + s_3 s_0$. The portfolio is designed such that each function class C_j has j local minima[1]. Individually, function C_4 is supposed to be more difficult to solve than C_1.

In Chook, instances are generated by (1) assigning a sub-function class to each subgraph in the problem, (2) a random rotation of the plaquette in the lattice. Instance classes are defined using a probability distribution over the sub-function classes: p_i the probability of selecting sub-functions from class C_i such that $\sum_i p_1 = 1$. Generating Tile Planting problems with Chook requires to set 4 parameters: the problem dimension n, and the probabilities p_1, p_2, and p_3. Perera *et al.* [16] analyze the performance of quantum algorithms (simulated quantum annealing) according to the main benchmark parameters p_i which can tune the problem difficulty. However in TP instances, the square toric lattice gives an identical role to binary variables. Moreover, the 2d shape of variables interactions allows the solving of such instances in polynomial time complexity [6].

---

[1] Indeed, j pairs of symmetric local minima.

3   PUBO$_i$ Problems

Here, we propose PUBO$_i$ for Polynomial Unconstrained Binary Optimization with importance, a generator in which benchmarks are expressed as Walsh functions, and takes the importance of variables into account. This version uses the same subproblems as in the Tile Planting generator, whose benchmarks correspond to Walsh functions of order 2. These subproblems are straightforward, which allows us to focus on our aim: to tune the variable importance to create more structured instances. Therefore, we focus on the introduction of three new parameters dedicated to this aim. In the following, we provide a mathematical description of Walsh functions. Then we present the first version of PUBO$_i$, as well as the benchmarks considered in our experiments.

3.1   PUBO Problems

Quadradratic Unconstrained Binary Optimization (QUBO), also known as Unconstrained Binary Quadratic Problem (UBQP) in the context of combinatorial optimization [13], are well known pseudo-boolean functions in the field of physics [4]. These functions are quadratic functions that can be generalized to any order. Although there exists several names, we denote this extension as PUBO for Polynomial Unconstrained Binary Optimization [7]. Notice that this extension is also known as spin glasses problems in physic. Several construction of PUBO exists, here we choose to present PUBO from the point of view of Walsh functions.

Given a bit string of dimension n, Walsh functions compose a finite set of 2^n pseudo-boolean functions defined for all integer k from $[0, 2^n - 1]$ by:

$$\begin{cases} \varphi_k : \{0,1\}^n \to \{-1,1\} \\ \quad x \mapsto (-1)^{\sum_{i=0}^{n-1} k_i x_i} \end{cases}$$

where x_i is the ith bit of x, and k_i is the ith bit representing the integer k. Walsh functions [25] is an orthonormal basis of pseudo-boolean functions. Any pseudo-boolean function $f : \{0,1\}^n \to \mathbb{R}$ can be written as: $f(x) = \sum_{k=0}^{2^n} w_k \varphi_k(x)$ with $w_k \in \mathbb{R}$. Indeed, Walsh transform is similar to Fourier transformation, and coefficients w_k are computed using the orthogonal projection for the L_2-norm of f on functions φ_k: $w_k = \frac{1}{2^n} \sum_{x \in \{0,1\}^n} f(x) \varphi_k(x)$.

Walsh functions allow a decomposition equivalent to a polynomial decomposition. Indeed, using the transformation of bit $x_i \in \{0,1\}$ into spin $s_i \in \{-1,1\}$ defined by $s_i = (-1)^{x_i}$, the kth Walsh function can be rewritten as $\varphi_k(s) = \prod_{i:k_i=1} s_i$. Thus, the order of a Walsh function φ_k is defined by the number of bits equal to 1 in the binary representation of k. For example, an order 2 Walsh function written as: $w_0^{(0)} + \sum_{i=0}^{n} w_i^{(1)} s_i + \sum_{i<j} w_{i,j}^{(2)} s_i s_j$.

For benchmarking purposes, in addition to being able to represent any evaluation function, Walsh functions allow to fine-tune both the interdependence between the variables (non-zero terms of the polynomial) and the intensity of the interactions ($|w_k|$ values). Using integer numbers for w_k, neutrality levels (plateaus) can also be tuned.

3.2   Definition of PUBO$_i$

The set of PUBO functions is a vector space of dimension 2^n where n is the bit string length. A benchmark based on PUBO has to define a subspace from this large vector space. The proposed benchmark PUBO$_i$ is defined with two principles. In order to reduce the dimension, following the TP benchmark design principle, PUBO$_i$ decomposes the objective function into a sum of sub-functions from a portfolio. In addition, the binary variables in each sub-function are selected according to tunable parameters of variable importance, to introduce a real-like property. As Tile Planting instances, PUBO$_i$ defines the objective as a sum of sub-functions:

$$\forall x \in \{0,1\}^n, \quad f(x) = \sum_{i=1}^{m} f_i(x)$$

where m is the number of sub-functions, and each sub-function f_i is selected at random according to probabilities p_j of each sub-function class C_j from the portfolio (see Sect. 2.2).

The main originality of the proposed benchmark lies around the notion of variable importance. Each sub-function depends on a limited number of variables. For example in TP instances, each sub-function depends on 4 variables selected according to the square lattices. In PUBO$_i$, variables are selected according to a degree of importance. Indeed, intuitively in real-world problems, important variables should appear more often in sub-problems than least important variables. To define the importance of variables, the set of variables $X = \{x_1, \ldots, x_n\}$ is split into k disjoint classes of importance: $c_i \subset X$ such that $\cup_k c_k = X$, and $c_i \cap c_j = \emptyset$ for each pair $\{i, j\}$. The number n_i of variables in each class is a parameter of the benchmark. Each class of importance c_i has a degree of importance $d_i \in \mathbb{R}^+$. In PUBO$_i$, the probability of selecting a variable in a sub-function is then proportional to the degree of variable importance of its class i: $p_{c_i} = \frac{d_i}{\sum_{j=1}^{k} d_j}$.

We also introduce another parameter to tune the co-appearance of variables in the same sub-function. In a real-world problem, one can assume important variables are not randomly distributed among the sub-problems. For some problem instances, important variables could appear together in the same sub-problems, and for other ones, important variables could be linked to less important ones. The following paragraph introduces the principle of co-appearance of important variables in PUBO$_i$.

For a sub-function of arity a, when classes of importance are independent, the probability of having variables of classes $c_{i_1}, \ldots, c_{i_a}$ is the product of probabilities $p_{c_{i_1}} \cdots p_{c_{i_a}}$. To tune this probability differently, the probability related to the number of each class is necessary. For simplification purposes, let us first suppose there are only 2 classes of importance (0 and 1). We denote $p_i^{(a)}$ the probability of having i variables of class 1 in the same sub-function of arity a. Then, we can define this probability recursively. For arity $a = 1$, $p_0^{(1)} + p_1^{(1)} = 1$. Thus, $p_0^{(1)} = p_{c_0}$, and $p_1^{(1)} = 1 - p_{c_0}$. The probability to select the class 0 for each sub-function variable should remain the same, hence the additional variable for arity

$a = 2$ should not change the marginal probability, and we have the following equations:

$$\begin{cases} p_0^{(2)} + p_1^{(2)} = p_0^{(1)} \\ p_1^{(2)} + p_2^{(2)} = p_1^{(1)} \end{cases}$$

The introduction of a new parameter p'_{co} to tune the different probabilities is possible. By setting $p_0^{(2)} = p'_{co} p_0^{(1)}$, we obtain:

$$\begin{cases} p_0^{(2)} = p'_{co} p_{co} \\ p_1^{(2)} = (1 - p'_{co}) p_{co} \\ p_2^{(2)} = (1 - p'_{co})(1 - p_{co}) + p'_{co} - p_{co} \end{cases}$$

When $p'_{co} = p_{co}$, the classes of importance of each variable are independent. When p'_{co} is greater than p_{co}, then the probability to have both variables in the same class of importance is higher. On the contrary, when p'_{co} is lower than p_{co}, the sub-functions have more heterogeneous classes of importance.

More generally, an additional variable in the sub-function leads to the recurrence formula: $\forall a \geq 2, \forall i \in \{0, \ldots, a\}$, $p_i^{(a)} + p_{i+1}^{(a)} = p_i^{(a-1)}$.

By setting the first probability $p_0^{(a)}$, one can deduce all other probabilities. It would be possible to introduce a parameter at each arity level: $p_0^{(a)} = p'^{(a)}_{co} p_0^{(a-1)}$. However, to simplify the benchmark tuning, we use one single parameter p'_{co}: $p_0^{(a)} = p'_{co} p_0^{(a-1)}$. Therefore we obtain the following values:

$$\begin{cases} p_0^{(a)} = (p'_{co})^{a-1} p_{co} \\ \ldots \\ p_i^{(a)} = (1 - p'_{co})^i (p'_{co})^{a-1-i} p_{co} \\ \ldots \\ p_a^{(a)} = (1 - p'_{co})^{a-1}(1 - p_{co}) + (1 - (1 - p'_{co})^{a-1})(1 - \frac{p_{co}}{p'_{co}}) \end{cases}$$

As for the arity 2, the value of p'_{co} determines the independence degree of co-appearance of the same class. Let us rewrite the parameter $p'_{co} = \alpha \, p_{co}$. When $\alpha = 1$, the co-appearance of variables importance classes are independent. When $\alpha > 1$, variables from the same class of importance have a higher probability to appear in the same sub-function, and conversely when $\alpha < 1$. Remember that globally on all sub-functions, each class of importance appears with a probability proportionally to its degree of importance. To extend to more than 2 classes of importance, we have to consider iteratively the probability to be in class 0, and not to be in class 0, then probability to be in class 1, and in a class greater than 1, etc.

We also propose to shift randomly the global minimum of each sub-function class. Naturally, knowing the global minimum could be useful, yet it would introduce a bias. Indeed, when a sub-problem is solved with the ground state 1111, it also helps the search to solve another sub-problem (shared variables are set to the optimal value). In PUBO_i, we propose to shift to a random binary string the minimum for each sub-function to design more challenging instances compared to those of Tile Planting. Of course, the knowledge of ground state is lost and should be approximated by the best known solution for each instance. However, a lower bound of minima can be computed by summing the minima value of each sub-function. Table 1 summarizes the parameters of PUBO$_i$ benchmark.

The code of the PUBO$_i$ generator is available on git https://gitlab.com/verel/pubo-importance-benchmark. The generator produces a file in json format which follows the same format of Chook generator. As a consequence, all solvers (quantum or classical) can be used to solve PUBO$_i$ instances.

Table 1. Parameters of PUBO$_i$ benchmark.

Parameter	Description	Experimental values
n	Problem dimension	$[1000, 5000]$
m	Number of sub-functions	$[0.01, 0.2] \times \frac{n(n-1)}{2}$
$\mathcal{C}$	Portfolio of sub-functions	Tile Planting
p_i	Probabilities of sub-function class	$[0, 1]$
k	Number of class of variable importance	2
n_i	Number of variables in each class of importance	$n_0 = 0.25n, \; n_1 = n - n_0$
d_i	Degree of importance of each class	$d_0 \in [1, 10], \; d_1 = 1$
α	Probability of importance class co-appearance	$[1, 1/(p_{c_0} - 1)]$

3.3   Instances Set

This instances set aims to provide a large set of PUBO$_i$ instances for the analysis of the benchmark parameters impact on the properties of the problems, but also to offer a diverse set instances to train, and test new efficient optimization algorithms for this type of problems, and real-world problems.

In this set of instances, we consider $k = 2$ classes of importance to distinguish important variables, and non-important variables. The number of important variables is set to 25% of the total number, and as a consequence, 75% of variables are considered non-important. The degree of importance of non-important variables is always set to $d_1 = 1$, while the degree of important variables is higher from the range $[1, 10]$. Important variables could be up to 10 times more frequent than non-important ones. The factor α is set to be larger than 1 up to the larger possible $\frac{1}{p_{c_0}-1}$. The problem dimension n is between 1000, and 5000 which is larger than in the TP set, and medium to large size according to the UBQP standard [13]. The number of sub-functions m is proportional to the square of problem dimension. Indeed, the portfolio of PUBO$_i$ is the one of TP which defines functions with 4 quadratic Walsh terms. Following the UBQP standard of matrix density, we choose m between 1% and 20% of $\frac{n(n-1)}{2}$. Table 1 summarizes the range of parameters value.

Instead of factorial design of experiments, we generate 1000 instances of $PUBO_i$ using *Latin hypercube sampling* [3] (LHS), a statistical method for the quasi-random sampling based on a multivariate probability distribution. Thus, in our context, LHS leads to a set of instances that covers a large panel of different parameter combination while significantly reducing the computational effort devoted to the study of such parameters. Indeed, the factorial design alternative either leads to a poorer coverage of the possible combinations or require the consideration of more instances, leading to a tedious process and a significantly higher computation cost. To respect the constraint $p_1 + p_2 + p_3 \leq 1$ to define, we reject samples from LHS which do not respect this constraint in order to avoid scaling bias. All instance files in json format are available online at the url https://gitlab.com/verel/pubo-importance-benchmark/instances.

4   Experimental Analysis

This set of experiments aims to highlight whether the parameters we propose impact variable importance, and how variable importance impacts some classic features of fitness landscapes.

4.1   Fitness Landscapes Characterization

The fitness landscape analysis mostly focuses on classic features: the ruggedness/multimodality and the neutrality. The ruggedness, or multimodality of a landscape mainly refers to the number of local optima, their distribution, and the size of their basins of attraction. This property reflects the difficulty to optimize the instance with local search algorithm based on given neighborhood relation. A *rugged* landscape has several peaks (local optima) hard to attain (small basins of attraction), whereas a *smooth* landscape has a few peaks easy to attain (large basins of attraction).

Here, we use a widely-used ruggedness indicator, the fitness *autocorrelation* [26]. Given a random walk $(x_t, x_{t+1}, \dots)$ where the solution x_{i+1} is a neighbor of x_i, the autocorrelation function ρ of the fitness function f corresponds to autocorrelation function of the time series $(f(x_t), f(x_{t+1}), \dots)$:

$$\rho(\ell) = \frac{E[f(s_t)f(s_{t+\ell})] - E[f(s_t)]E[f(s_{t+\ell})]}{var[f(s_t)]}$$

where $E[f(s_t)]$ is the expected value of $f(s_t)$, and $var[f(s_t)]$ its variance. In the following, we only consider the first coeffficient of autocorrelation $\rho(1)$, as it usually is enough to summarize the ruggeness levels of landscapes.

The neutrality of a fitness landscape refers to the proportion of neutral neighbors in the landscape. A neutral neighbor of a solution has the same fitness value. Neutrality often induces plateaus in the landscape, in which optimization methods can easily be trapped, preventing them to attain better-quality solutions. Thus, a fitness landscape with high neutrality levels can be challenging to optimize. We use the neutrality degree [19] on neutral moves performed during the random walks $(x_t, x_{t+1}, \ldots)$: $deg_n(x) = \#\{f(x_{i+1}) = f(x_i) : i \in \{0, 1, \ldots\}\}$. More precisely, we report the neutral mutation probability $\frac{deg_n(x)}{n}$ which is independent of problem dimension.

The last measure considered for the fitness landscape analysis is the length of the adaptive walk. An adaptive walk is a sequence of neighboring and improving solutions: $(x_0, x_1, \ldots, x_\ell)$ such for all $i \in \{0, \ldots, \ell\}$, x_{i+1} is a neighbor of x_i, and $f(x_{i+1})$ is better than $f(x_i)$. Such a measure highlights the multimodality (i.e., the presence of several peaks) of fitness landscapes. A short walk is generally the sign of a multimodal landscape, whereas a long walk usually indicates a monomodal landscape. Several adaptive walks exist, and on combinatorial fitness landscape hill-climbers are often considered to this aim. Here, we use a first improvement hill-climber which consists of randomly selecting an improving neighbor at each step of the search, until a local optimum is met. The length of adaptive walk is then the number of steps ℓ.

4.2   Methodology

We consider the set of 1000 instances generated with LHS. For each instance, we conducted a run of random walk of length $30n$. This length is sufficient to provide a sample of solutions, as each variable is flipped 30 times on average, and allows to compute statistics for each variable. We consider a classic neighborhood-relation of pseudo-boolean optimization problems, 1-flip. 30 adaptive walks are conducted for each landscape, starting from a randomly generated solution, and stopping on a local optimum. The average length of adaptive walks is reported. As our goal is to observe the impact of benchmark parameters, and in particular variable importance on landscape characteristics, we use the mgvc R package [27] to generate *generalized additive models* (GAMs) to highlight the significance of these parameters on the variable importance. GAMs are statistical models that merge the properties of the generalized linear model with those of the additive model. These results are presented through scatterplots of the regression model between landscape features and benchmark parameters.

4.3   Results

Figure 3 shows the scatter plot with the GAM model regression of landscape features, autocorrelation coefficient, the neutral mutation probability, and the length of adaptive walks (see Sect. 4.1), according to the benchmark parameters. Each plot reports the *effective degrees of freedom* (edf) of GAM, which reflects the degree of non linearity: $edf = 1$ corresponds to a linear relationship, $1 < edf \leq 2$ is a quadratic, or weakly non-linear relationship and $edf > 2$ a highly non-linear relationship [29]. The significance values (Signif.) are also reported. The asterisks indicates the p-value at different levels of the statistical significance: '***', '**', '*', '.', and '-' correspond respectively to p-value levels of 0, 0.001, 0.01, 0.1, 0.05, and 1.

The problem dimension (n), the number of sub-functions (m), the degree of importance, and the probability p_1 of choosing the sub-problem C_1 have a

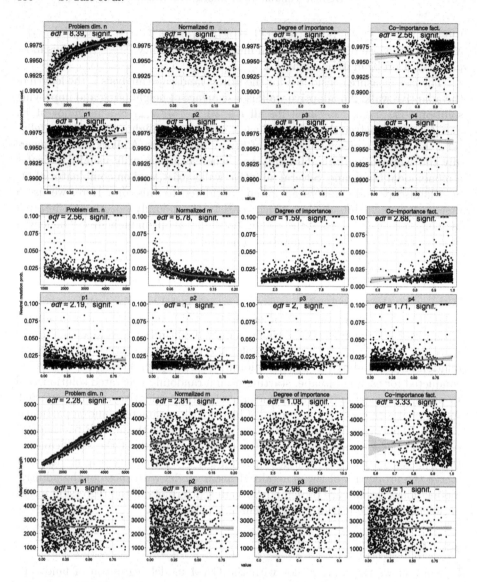

Fig. 3. Scatterlots features *vs.* benchmark parameters with GAM model regression.

highly significant impact on the autocorrelation (ruggedness) of landscapes. The
factor of independence α has a significant impact on the autocorrelation. The
autocorrelation increases, meaning ruggedness level decreases, with the problem
dimension, and the factor α. The relation between autocorrelation, and problem
dimension is a highly non-linear relationship. Ruggedness levels also decrease
with the degree of importance d_0, and the presence of subproblem C_1. For these
two parameters, the relationship with the autocorrelation is linear. With larger

values of m, the ruggedness level increases linearly. Higher ruggedness levels are coherent with a higher number of subproblems as it increases the complexity of the problem, and the interdependence between subproblems. n has the highest impact on the autocorrelation, which is consistent with general results on landscapes. Specific parameters to PUBO$_i$ on variable importance also impacts ruggedness levels, among those α has the highest impact.

Another facet of problem difficulty is the neutrality feature. All benchmark parameters n, m, d_0, α and p_4 (except p_1, p_2, and p_3) have a significant impact on neutrality level. While increasing both problem dimension and number of sub-functions decrease neutrality rates, higher values of the importance degree d_0, factor α and p_4 increase these rates. Parameters m and α have the highest effect on neutrality of the problem instances.

The impact of benchmark parameters on the length of adaptive walk is usually lower than for the autocorrelation and neutrality rates. The problem dimension significantly increases this length, as on most combinatorial problems. A larger number of subproblems seems to increase the multimodality of landscapes. Nonetheless, the walk length does not take into account the variation of fitness between two steps of adaptive walks, limiting further interpretations.

The autocorrelation coefficient, and probability of neutral mutation can be computed individually for each variable. This allows to study the impact of benchmark parameters on each importance class of variables. Figure 4 shows the scatter plot with GAM model regression of the landscape features difference between important and non-important variables *versus* the benchmark parameters. Both for the autocorrelation and neutrality rates, values are negative, meaning that the contribution of important variables is lower than the one of non-important variables: the subspace of important variables is more rugged, and less neutral (flat) than the subspace of non-important variables. With the increase of n, and m, the impact of important variables on the autocorrelation significantly decreases, with average values respectively ranging from -0.012 to -0.003, and 0.0075 to 0.004. The opposite happens with the importance degree d_0, where the average difference of autocorrelation values significantly increases from 0.0025 to 0.007. Increasing the importance degree significantly increases the ruggedness contribution of important variables. The factor of independence α has a significant impact on the contribution of important variables to the autocorrelation. When this factor is increased by a small gap, important variable contribute more to the ruggedness. Interestingly, while ruggedness level is low, it seems higher around important variables, possibly indicating these landscapes are globally smooth but locally rugged, as some UBQP landscapes [21].

Except for the subproblems C_i, PUBO$_i$ parameters strongly impact the neutrality level of important variables. Increasing the degree of importance d_0 and the factor α raise the contribution of important variables on neutrality rates. On larger instances and instances with more subproblems, the contribution of important variables wanes while remaining higher than the one of non-important variables. Note that some portfolio subproblems have a higher impact on the difference for neutrality levels than ruggedness, indeed C_1 and C_4 contain 3 out of 4 same contribution values.

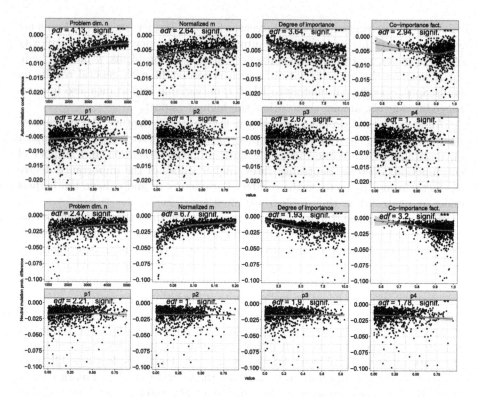

Fig. 4. Scatterlots of features difference between important and non-important variables *vs.* benchmark parameters with GAM model regression.

5 Discussion and Future Work

The experimental analysis shows that except for parameters which tune the proportion of subproblem classes (p_i), all of PUBO$_i$ parameters have a significant impact on ruggedness, multimodality and neutrality levels of landscapes. In particular, parameters related to variable importance could have the same impact on landscape, and so on the problem difficulty, than classical parameter such as problem dimension. Moreover, the tunable importance of the benchmark leads to non isotropic landscapes where the features of landscapes are different for the subspace of important variables. To our best knowledge, this property of importance is rarely taking into account in benchmark design. The difference of landscape features between important and non-important variables shows that this property should be considered in the design of evolutionary, and local search algorithms. In particular, the design of local search operator, and neighborhood should be designed according to the variable importance either by an expert, using machine learning technique, or the both. The PUBO$_i$ generator allows us to facilitate this approach by bringing a large set of diverse instances.

One natural perspective is to able to compare real-world combinatorial problems to PUBO$_i$ instances, in particular according to the variable importance property. It would also be relevant to study other possible benchmark parameters such as the number of importance classes, and more deeply the composition of portfolio. Although we have been able to analyze the fitness landscapes of PUBO$_i$ instances, new analysis should be conducted using for example Local Optima Network. Moreover, new fitness landscape analysis tools should be designed in order to sharply describe the anisotropy of a landscape. Of course, the goal of this benchmark is to train, test, and understand new optimization algorithms (quantum or classical), and further developments could be conducted in this research direction. The design methodology of variable importance used in PUBO$_i$ is generic, and another considered research direction is to extend the generator to other type of optimization problems (continuous, etc.).

Acknowledgements. Experiments presented in this paper were carried out using the CALCULCO computing platform, supported by SCoSI/ULCO (Service COmmun du Système d'Information de l'Université du Littoral Côte d'Opale).

References

1. Bartz-Beielstein, T., et al.: Benchmarking in optimization: Best practice and open issues. arXiv preprint arXiv:2007.03488 (2020)
2. Burkard, R.E., Karisch, S.E., Rendl, F.: QAPLIB-a quadratic assignment problem library. J. Global Optim. **10**(4), 391–403 (1997)
3. Carnell, R.: LHS: Latin hypercube samples, R package version 0.16 (2018)
4. Date, P., Patton, R., Schuman, C., Potok, T.: Efficiently embedding QUBO problems on adiabatic quantum computers. Quantum Inf. Process. **18**(4), 1–31 (2019). https://doi.org/10.1007/s11128-019-2236-3
5. Doerr, C., Ye, F., Horesh, N., Wang, H., Shir, O.M., Bäck, T.: Benchmarking discrete optimization heuristics with IOH profiler. Appl. Soft Comput. **88**, 106207 (2020)
6. Galluccio, A., Loebl, M., Vondrák, J.: Optimization via enumeration: a new algorithm for the max cut problem. Math. Program. **90**(2), 273–290 (2001)
7. Glover, F., Hao, J.K., Kochenberger, G.: Polynomial unconstrained binary optimisation-part 1. Int. J. Metaheuristics **1**(3), 232–256 (2011)
8. Hamze, F., Jacob, D.C., Ochoa, A.J., Perera, D., Wang, W., Katzgraber, H.G.: From near to eternity: spin-glass planting, tiling puzzles, and constraint-satisfaction problems. Phys. Rev. E **97**(4), 043303 (2018)
9. Hansen, N., Auger, A., Ros, R., Finck, S., Pošík, P.: Comparing results of 31 algorithms from the black-box optimization benchmarking BBOB-2009. In: GECCO, pp. 1689–1696 (2010)
10. Hansen, N., Auger, A., Ros, R., Mersmann, O., Tušar, T., Brockhoff, D.: COCO: A platform for comparing continuous optimizers in a black-box setting. Optim. Methods Softw. **36**(1), 114–144 (2021)
11. Johnson, D.S., Trick, M.A.: Cliques, coloring, and satisfiability: second DIMACS implementation challenge, 11–13 October, 1993, vol. 26. Am. Math. Soc. (1996)
12. Leprêtre, F., et al.: Fitness landscapes analysis and adaptive algorithms design for traffic lights optimization on SIALAC benchmark. Appl. Soft Comput. **85**, 105869 (2019)

13. Lü, Z., Glover, F., Hao, J.K.: A hybrid metaheuristic approach to solving the UBQP problem. Eur. J. Oper. Res. **207**(3), 1254–1262 (2010)
14. Malan, K.M., Engelbrecht, A.P.: A survey of techniques for characterising fitness landscapes and some possible ways forward. Inf. Sci. **241**, 148–163 (2013)
15. Malan, K.M.: A survey of advances in landscape analysis for optimisation. Algorithms **14**(2), 40 (2021)
16. Perera, D., et al.: Chook-a comprehensive suite for generating binary optimization problems with planted solutions. arXiv preprint arXiv:2005.14344 (2020)
17. Sagawa, M., et al.: Learning variable importance to guide recombination. In: 2016 IEEE SSCI, pp. 1–7 (2016)
18. Sagawa, M., et al.: Learning variable importance to guide recombination on many-objective optimization. In: 6th IIAI-AAI, pp. 874–879. IEEE (2017)
19. Schuster, P., Fontana, W., Stadler, P.F., Hofacker, I.L.: From sequences to shapes and back: a case study in RNA secondary structures. Proc. R. Soc. London. Ser. B. Biol. Sci. **255**(1344), 279–284 (1994)
20. Taillard, E.: Benchmarks for basic scheduling problems. Eur. J. Oper. Res. **64**(2), 278–285 (1993)
21. Tari, S., Basseur, M., Goëffon, A.: Sampled walk and binary fitness landscapes exploration. In: Lutton, E., Legrand, P., Parrend, P., Monmarché, N., Schoenauer, M. (eds.) EA 2017. LNCS, vol. 10764, pp. 47–57. Springer, Cham (2018). https://doi.org/10.1007/978-3-319-78133-4_4
22. Tušar, T., Brockhoff, D., Hansen, N.: Mixed-integer benchmark problems for single- and bi-objective optimization. In: Proceedings of the Genetic and Evolutionary Computation Conference, pp. 718–726 (2019)
23. Tušar, T., Brockhoff, D., Hansen, N., Auger, A.: COCO: the bi-objective black box optimization benchmarking (bbob-biobj) test suite. ArXiv e-prints (2016)
24. Verel, S., Derbel, B., Liefooghe, A., Aguirre, H., Tanaka, K.: A surrogate model based on Walsh decomposition for pseudo-Boolean functions. In: Auger, A., Fonseca, C.M., Lourenço, N., Machado, P., Paquete, L., Whitley, D. (eds.) PPSN 2018. LNCS, vol. 11102, pp. 181–193. Springer, Cham (2018). https://doi.org/10.1007/978-3-319-99259-4_15
25. Walsh, J.L.: A closed set of normal orthogonal functions. Am. J. Math. **45**(1), 5 (1923)
26. Weinberger, E.: Correlated and uncorrelated fitness landscapes and how to tell the difference. Biol. Cybern. **63**(5), 325–336 (1990)
27. Wood, S.: MGCV: GAMs in R. Generalized Additive Mixed Models Using MGCV and LME4 (2012)
28. Wright, S.: The roles of mutation, inbreeding, crossbreeding, and selection in evolution. In: Proceedings of the Sixth International Congress of Genetics, vol. 1, pp. 356–366 (1932)
29. Zuur, A.F., Ieno, E.N., Walker, N.J., Saveliev, A.A., Smith, G.M.: Things are not always linear; additive modelling. In: Mixed effects models and extensions in ecology with R. SBH, pp. 35–69. Springer, New York (2009). https://doi.org/10.1007/978-0-387-87458-6_3

Stagnation Detection Meets Fast Mutation

Benjamin Doerr[1(✉)] and Amirhossein Rajabi[2(✉)]

[1] Laboratoire d'Informatique (LIX), CNRS, École Polytechnique Institute Polytechnique de Paris, Palaiseau, France
[2] Technical University of Denmark, Kgs. Lyngby, Denmark
amraj@dtu.dk

Abstract. Two mechanisms have recently been proposed that can significantly speed up finding distant improving solutions via mutation, namely using a random mutation rate drawn from a heavy-tailed distribution ("fast mutation", Doerr et al. (2017)) and increasing the mutation strength based on a stagnation detection mechanism (Rajabi and Witt (2020)). Whereas the latter can obtain the asymptotically best probability of finding a single desired solution in a given distance, the former is more robust and performs much better when many improving solutions in some distance exist.

In this work, we propose a mutation strategy that combines ideas of both mechanisms. We show that it can also obtain the best possible probability of finding a single distant solution. However, when several improving solutions exist, it can outperform both the stagnation-detection approach and fast mutation. The new operator is more than an interleaving of the two previous mechanisms and it outperforms any such interleaving.

Keywords: Mutation operator · parameter control · jump functions · theory

1 Introduction

Leaving local optima is a challenge for evolutionary algorithms. Mutation-based approaches are challenged by the fact that the typical mutation rate of $p = 1/n$ rarely leads to offspring in a larger distance from the parent. When using larger mutation rates, the choice of the mutation rate is critical and small constant-factor deviations from the optimal rate can lead to huge performance losses [11, Cor. 4.2].

Two ways to overcome this problem were proposed recently, namely the use of a random mutation rate sampled from a power-law distribution ("fast mutation") [11] and the successive increase of the mutation rate when a stagnation-detection mechanism indicates that the current rate is unlikely to generate solutions not seen yet [19]. An improved version of this stagnation-detection approach [21], the so-called SD-RLS algorithm based on k-bit mutation instead of standard bit mutation, can find a single improving solution in distance m in expected time $(1 + o(1))\binom{n}{m}$ (without knowing that the distance to the desired solution is m). Apart from lower order terms, this is the same runtime that can be obtained via a repeated use of the best unbiased mutation operator that is aware of m (which is, naturally, flipping m random bits). It is faster than the fast $(1 + 1)$ EA by a factor of $\Omega(m)$.

While the SD-RLS algorithm thus is very efficient in finding a single desired solution (and thus has very good runtimes on the classic jump functions benchmark (see Sect. 5 for a definition)), this algorithm has a poor performance when there are several improving solutions in distance m as now the stagnation detection approach leads to too much time spent on too small mutation strengths. Taking as an extreme example the generalized jump function [5] (see again Sect. 5 for a definition) having a valley of low fitness of width δ, $\delta \geq 2$ a constant, in distance $n/4$ from the optimum, we easily see that the SD-RLS takes an expected time of $\Omega(n^{\delta-1})$ to traverse the fitness valley, whereas the $(1 + 1)$ EA both with the classic mutation operator and with fast mutation does so in expected constant time.

Our Results: Based on the insight that fast mutation and stagnation detection have complementary strengths, we design a mutation-based approach that takes inspiration from both approaches. We follow, in principle, the basic version of the improved stagnation-detection approach of [21], that is, we start with mutation strength $r = 1$ and increase r gradually. More precisely, when strength r has been used for a certain number ℓ_r of iterations without that an improvement was found, we increase r by one since we assume that no improvement in distance r exists (we omit some technical details in this first presentation of our approach, e.g., that we do not increase r beyond $n/2.1$, and refer the reader to Algorithm 1 for the full details). Different from [21], when the current strength is r, we do not always flip r random bits as mutation operation, but we choose a random number X_r of bits to flip. This number is equal to r with probability $1 - \gamma$, where γ is an algorithm parameter that is usually small (a small constant or $o(1)$). With probability γ, however, X_r deviates from r by an amount following a power-law distribution with exponent β. The precise definition of this case (see again Algorithm 1) is not too important, so for this first exposition we can assume that we sample D from a power-law distribution (with exponent β) on the positive integers and then, each with probability $1/2$, flip $r + D$ or $r - D$ random bits (where we do nothing if this number is not between 1 and n).

Since with probability $1 - \gamma$ we essentially follow the basic approach of [21], it is not surprising that we find a single closest improving solution in distance m in an expected time of $\frac{1}{1-\gamma}(1 + o(1))\binom{n}{m}$, again without that the algorithm needs to know m (Theorem 5). If $\gamma = o(1)$, this is again the optimal time of $(1 + o(1))\binom{n}{m}$ discussed above. We note, however, that our algorithm is simpler than the solution presented in [21]. The basic SD-RLS algorithm proposed in [21] obtains a runtime of $(1+o(1))\binom{n}{m}$ only with high probability and otherwise fails. To turn this algorithm into one that never fails and has an expected runtime of $(1+o(1))\binom{n}{m}$, a robust version of the SD-RLS was developed in [21] as well. This version repeats previous phases as follows. When the ℓ_r uses of strength r have not led to an improvement, before increasing the rate to $r + 1$, first another ℓ_i iterations are performed with strength i, for $i = r - 1, \ldots, 1$. In our approach, such an additional effort is not necessary since the fast mutations automatically render the algorithm robust.

The use of a heavy-tailed mutation rate also helps in situations where the stagnation-detection mechanism takes too long to use larger mutation strengths. Since in phases $r = 1, \ldots, 2m$ the probability to flip m bits is at least $\gamma/2$ times the probability of this event in a run of the fast $(1 + 1)$ EA, it is not surprising that our algorithm finds an improvement in distance m is at most $2/\gamma$ times the time of the fast $(1 + 1)$ EA, which as discussed above can be significantly faster than the SD-RLS. Such a result could also have been obtained from a simple interleaving of SD-RLS and fast $(1 + 1)$ EA iterations. Since our heavy-tailed choices of the mutation strength, however, take into account the current strength r, we often obtain better runtimes, often better than both the SD-RLS and the fast $(1 + 1)$ EA. As the precise statement of these results is technical, we defer the details to Sect. 4.

As a simple example showing the outperformance of our algorithm, we regard the generalized jump function $\text{JUMP}_{m,\delta:=m-\Delta}$ for a constant value of $\Delta \geq 2$ and $m = \omega(1)$. This jump function is similar to the classic jump function JUMP_m, but the valley of low fitness consists not of all search points in positive distance at most $m - 1$ from the optimum, but only of those in distance $\Delta + 1, \ldots, m - 1$. Consequently, from the local optimum there is not a single improving solution, but $\Theta(n^\Delta)$. Note that this is still relatively few compared to the fitness valley of size essentially $\binom{n}{m-1}$. On this generalized jump function, the expected runtime of SD-RLS is $O\left(\binom{n}{\delta-1}\ln(R)\right)$, the one of the fast $(1 + 1)$ EA is $O\left(\delta^{\beta-0.5}(en/\delta)^\delta n^{-\Delta}\right)$, and the one of our algorithm is at most $O\left(\binom{n}{\delta}n^{-\Delta}\gamma^{-1}\right)$ (Corollary 11). Since it is also clear that any interleaving of SD-RLS and fast $(1 + 1)$ EA iterations cannot give a better runtime than the one of the two pure algorithms, this result shows that our algorithm can beat SD-RLS and fast EA (and any simple mix of them) when there are several improving solutions in a given distance.

A short experimental evaluation of the algorithms discussed so far shows that the advantages of our algorithm, proven only via asymptotic runtime results, are also visible for moderate problems sizes.

Structure of This Paper: After reviewing the most relevant previous works in Sect. 2, we introduce our new algorithm in Sect. 3. In Sect. 4, we analyze via mathematical means how our algorithm finds an improvement in distance m both when this is typically achieved in phase m (e.g., when there is only one improving solution in distance m) and when this is achieved earlier via the heavy-tailed rates. We use these results in Sect. 5 to prove several runtime results, among others, for generalized jump functions. We present some experimental results in Sect. 6. In Sect. 7, we discuss recommendations on how to set the parameters of our algorithm. We conclude the paper with a short discussion of our results and a pointer to possible future work in Sect. 8. Due to space restrictions, all proofs had to be omitted from this paper; however, they are available in the preprint [12].

2   Previous Works

This work aims at combining the advantages of stagnation detection and heavy-tailed mutation, so clearly these topics contain the most relevant previous works. Both integrate into the wider questions of how to optimally set the mutation strength of evolutionary algorithms (for this we refer to the recent survey [10]) and how evolutionary algorithms can leave local optima (here we refer to [9, Section 2.1] for a discussion of non-elitist approaches and to the introduction of [8] for a discussion of crossover-based approaches).

For elitist mutation-based approaches, it is clear that when the population has converged to a local optimum the only way to leave this is by mutating a solution from the local optimum into an at least as good solution outside this local optimum. It was observed in [11] (the earlier work [18] contains similar findings for the special case that the nearest improving solution is in Hamming distance two or three) that standard bit mutation with mutation rate $p = \frac{1}{n}$, which is the most recommended way of doing mutation, is not perfectly suitable to perform larger jumps in the search space. In fact, when the nearest improving solution is in Hamming distance m, then a mutation rate of $p = \frac{m}{n}$ is much better, leading to a speed-up by a factor of order $m^{\Theta(m)}$.

Since [11] also observed that missing the optimal rate by a small constant factor leads to performance losses exponential in m, it was proposed to use a mutation rate that is drawn from a (heavy-tailed) power-law distribution. Without the need to know m, this approach led to runtimes that exceed the ones obtained from the optimal rate $p = \frac{m}{n}$ by only a small factor polynomial in m. This price for universality can be made as low as $\Theta(m^{0.5+\varepsilon})$, but not smaller than $\Theta(\sqrt{m})$. Various variants of heavy-tailed mutation operators have been proposed subsequently, also heavy-tailed choices of other parameters have been used with great success [1–4, 6, 7, 13, 15–17, 24].

A different way to cope with local optima was proposed in [19]. When an algorithm is stuck in a local optimum for a sufficiently long time, then with high probability it has explored all search points in a certain radius. Consequently, it is safe to increase the mutation rate, which increases the probability to generate more distant solutions. This is the main idea of a series of works on stagnation detection [19–21]. As shown in [19], this approach can save the polynomial price for universality of the heavy-tailed approach and thus obtain runtimes of the same asymptotic order as when using the optimal (problem-specific) mutation rate. By replacing standard bit mutation with m-bit flips, the time to find a particular solution in Hamming distance m was further reduced to $(1+o(1))\binom{n}{m}$, the same time (apart from lower order terms) one would obtain with the best unbiased mutation operator (which consists of flipping m random bits).

To be precise, two approaches are discussed in [21]. The simple one, obtained from just replacing standard bit mutation in [19] by r-bit mutation, obtains the desired runtimes with high probability, but fails completely with some very small probability. For this reason, also a robust version of the algorithm was proposed in [21], which by cyclically reverting to smaller mutation strengths overcomes the problem that, with small probability, a given solution in distance m is not

---

Algorithm 1: The SD-FEA$_{\beta,\gamma,R}$ for the maximization of $f \colon \{0,1\}^n \to \mathbb{R}$. Its parameters are the power-law exponent $\beta > 1$, the probability γ to deviate from rate r in phase r, and the parameter R which defines the maximum length of the r-th phase at $\ell_r = (1 - \gamma)^{-1} \binom{n}{r} \ln(R)$.

---

1 Select x uniformly at random from $\{0,1\}^n$ and set $r_1 \leftarrow 1$;
2 $u \leftarrow 0$;
3 **for** $t \leftarrow 1, 2, \ldots$ **do**

4 $\quad$ Set $s = r_t$ $\qquad\qquad\qquad\qquad$ with probability $1-\gamma$ or
    $\quad\quad s = r_t + \mathrm{pow}(\beta, n - r_t)$ $\qquad$ with probability $\gamma/2$ or
    $\quad\quad s = r_t - \mathrm{pow}(\beta, \max\{1, r_t-1\})$ with probability $\gamma/2$;
5 $\quad$ Create y by flipping s bits in a copy of x uniformly at random;
6 $\quad$ $u \leftarrow u + 1$;
7 $\quad$ **if** $f(y) > f(x)$ **then**
8 $\quad\quad$ $x \leftarrow y$;
9 $\quad\quad$ $r_{t+1} \leftarrow 1$;
10 $\quad\quad$ $u \leftarrow 0$;
11 $\quad$ **else if** $f(y) = f(x)$ and $r_t = 1$ **then**
12 $\quad\quad$ $x \leftarrow y$;
13 $\quad$ **if** $u \geq \ell_{r_t}$ **then**
14 $\quad\quad$ $r_{t+1} \leftarrow \min\{r_t + 1, \lfloor \frac{n}{2.1} \rfloor\}$;
15 $\quad\quad$ $u \leftarrow 0$;
16 $\quad$ **else**
17 $\quad\quad$ $r_{t+1} \leftarrow r_t$;

---

found in the phase which uses m-bit flips. In [20], a variation of SD-RLS was proposed that keeps the successful strength after leaving local optima with the help of the radius memory mechanism, which is beneficial on highly multimodal fitness landscapes. The idea of stagnation detection has also been successfully used in multi-objective evolutionary computation [13].

3 Combining Fast Mutation and Stagnation Detection: The Algorithm SD-FEA$_{\beta,\gamma,R}$

We propose the algorithm SD-FEA$_{\beta,\gamma,R}$ for the maximization of pseudo-Boolean functions $f \colon \{0,1\}^n \to \mathbb{R}$ as defined in Algorithm 1. The function $\mathrm{pow}(\beta, u)$ samples from a power-law distribution with exponent β and range $[1..u]$ as defined in Eq. (1) below.

The general idea of this algorithm is that it increases the mutation strength r to $r + 1$ when the improvement is not in Hamming distance r with at least a constant probability (with probability $1/R$ roughly) using the stagnation detection mechanism. While the strength is r, called in phase r, the algorithm looks at larger or smaller Hamming distances (with probability γ) besides using the current strength r. The distribution of the distance of the search radius from to the current strength r follows a power-law distribution. An integer random variable X follows a power-law distribution with parameters β and u if

$$\Pr[X = i] = \begin{cases} C_{\beta,u} i^{-\beta} & \text{if } 1 \leq i \leq u, \\ 0 & \text{otherwise,} \end{cases} \tag{1}$$

where $C_{\beta,u} := (\sum_{j=1}^{u} j^{-\beta})^{-1}$ is the normalization coefficient. The function $\mathrm{pow}(\beta, u)$ used in Algorithm 1 returns a sample from this distribution.

The algorithm starts with a search point selected uniformly at random from the search space $\{0,1\}^n$ and with the initial strength $r = 1$. There is a counter u for counting the number of unsuccessful steps in finding a strict improvement with the current strength. When the counter exceeds the maximum phase length ℓ_r, the strength r increases by one but not exceeding $n/2.1$. When the algorithm makes progress, the counter and strength are reset to their initial values.

The mutation, which we call s-flip in the following, flips exactly s bits randomly chosen as follows. With probability $1-\gamma$, the algorithm flips exactly r bits in phase r. However, with probability γ, the algorithm deviates from this choice and instead flips a number of bits which differs from r, in either direction, by a value following a power-law distribution. The distribution over s is analyzed in Lemma 1 below.

In this paper, we use maximum phase lengths of

$$\ell_r = \binom{n}{r} / (1 - \gamma) \ln(R). \tag{2}$$

This choice is designed for pseudo-Boolean fitness functions. For other search spaces, the maximum phase length should be $\ell_r = |S_r|/(1 - \gamma) \ln(R)$, where $|S_r|$ is the number of search points in distance r from the current search point or an upper bound for this. The maximum phase length defined in Eq. (2) has a parameter R controlling the probability of failing to find an improvement at the "right" strength. To prove our theoretical results, R should be selected at least $e^{1/\gamma}$. In Sect. 7, we give some recommendations for choosing the parameters of the SD-FEA$_{\beta,\gamma,R}$.

As *runtime* of a heuristic algorithm on a fitness function f, we define the first point of time t where a search point of maximal fitness has been evaluated.

4   Analysis of the SD-FEA$_{\beta,\gamma,R}$

In this paper, let us define by the *individual gap* of $x \in \{0,1\}^n$ the minimum Hamming distance of x from points with strictly larger fitness function value, that is,

$$\text{IndividualGap}(x) := \min\{H(x,y) : f(y) > f(x), y \in \{0,1\}^n\}.$$

By the *fitness level* of x, we mean all the search points with fitness value $f(x)$. We call the *fitness level gap* of a point $x \in \{0,1\}^n$ the maximum of all individual gap sizes in the fitness level of x, i. e.,

$$\text{FitnessLevelGap}(x) := \max\{\text{IndividualGap}(y) : f(y) = f(x), y \in \{0,1\}^n\}.$$

If the algorithm creates a point at the Hamming distance IndividualGap(x) from the current search point x, with positive probability an improvement can be found. Note that FitnessLevelGap(x) = 1 is allowed, so the definition also covers search points that are not local optima. As long as a strict improvement is not made, the FitnessLevelGap remains the same, although the current search point might be replaced with another search point in the fitness level in phase 1, that is, when the strength is 1.

We now analyze how the SD-FEA$_{\beta,\gamma,R}$ finds better selections. Let the current search point be x. We define by phase r all points of time where radius r is used for search points with fitness value $f(x)$, i.e., while in the fitness level of x. Let E_r be the event of **not** finding the optimum within phase r. For $j \geq i$, let E_i^j denote the event of not finding a strict improvement within phases i to j. Formally, $E_i^j = E_i \cap \cdots \cap E_j$.

Before computing the probabilities of these events, we need to know the distribution of the offspring in an iteration. The following lemma will be used throughout this paper, showing the distribution of the number of flipping bits (i.e., the variable s in Algorithm 1) in each iteration. We recall that in phase r, with a relatively large probability $1 - \gamma$, the algorithm flips r bits. However, with probability γ, it uses power-law distributions to flip less or more than r bits.

Lemma 1. *Let r be the current strength in an iteration of the algorithm SD-FEA$_{\beta,\gamma,R}$. Let X be the integer random variable corresponding to the number of bits that are flipped, that is, the variable s in Algorithm 1. Then*

$$\Pr[X = \alpha] = \begin{cases} (\gamma/2) \cdot C_{\beta,r-1} \cdot (r - \alpha)^{-\beta} & 1 \leq \alpha < r, \\ 1 - \gamma & \alpha = r, \\ (\gamma/2) \cdot C_{\beta,n-r} \cdot (\alpha - r)^{-\beta} & r < \alpha \leq n, \end{cases}$$

and for $r = 1$, $\Pr[X = 0] = \gamma/2$.

The following lemma estimates the probability of reaching a phase that is greater than the fitness gap size. In the statement of the lemma, recall that the parameter R controls the length of the phase.

Lemma 2. *Let $\beta > 1$, $0 < \gamma < 1$ and $R > 1$. Consider the SD-FEA$_{\beta,\gamma,R}$ maximizing a pseudo-Boolean fitness function $f: \{0,1\}^n \to \mathbb{R}$. Let $x \in \{0,1\}^n$ be the current search point immediately following a strict improvement or the initial search point. Let $m = \text{IndividualGap}(x)$. Let E_1^{r-1} denote the probability of not finding an improvement in phases 1 to $r - 1$. Then for $m < r \leq \lfloor \frac{n}{2.1} \rfloor$, we have*

$$\Pr[E_1^{r-1}] \leq R^{-1-(\gamma/2)\cdot\left(\frac{\ln(1.1)}{\beta}\right)^{\beta} C_{\beta,n}(r-m-1)}.$$

The next lemma is used to estimate the number of iterations in phases larger than the fitness level gap. With a good choice of the parameters γ and R, the following result becomes $o(1/s_m)$, that is, the number of steps at larger strengths is negligible compared to the number of steps at the phase m.

Lemma 3. *Let $\beta > 1$, $0 < \gamma < 1$ and $R \geq e^{1/\gamma}$. Consider the SD-FEA$_{\beta,\gamma,R}$ maximizing a pseudo-Boolean fitness function $f : \{0,1\}^n \to \mathbb{R}$. Let $x \in \{0,1\}^n$ be the current search point immediately following a strict improvement or the initial search point. Assume $m = \text{FitnessLevelGap}(x)$ and $m \leq \lfloor n/2.1 \rfloor$. Let s_m be a lower bound on the probability that an improvement is found from search points in the fitness level of x conditional on flipping m bits. Then the expected number of iterations spent with strengths larger than m is at most*

$$O\left(R^{-1}\gamma^{-1}\frac{1}{s_m}\right).$$

The following lemma, a combinatorial inequality taken from [21], will be used to count the number of iterations spent with strengths smaller than the fitness level gap.

Lemma 4 (Lemma 1 in [21]). *For any integer $m \leq n/2$, we have*

$$\sum_{i=1}^{m}\binom{n}{i} \leq \frac{n-(m-1)}{n-(2m-1)}\binom{n}{m}.$$

We now present the first main result. In the following theorem, we provide two rigorous upper bounds on the escaping time from a local optimum.

Theorem 5. *Let $\beta > 1$, $0 < \gamma < 1$ and $R \geq e^{1/\gamma}$. Consider the SD-FEA$_{\beta,\gamma,R}$ maximizing a pseudo-Boolean fitness function $f : \{0,1\}^n \to \mathbb{R}$. Let $x \in \{0,1\}^n$ be the current search point immediately following a strict improvement or the initial search point. Let $m = \text{FitnessLevelGap}(x)$. Define T as the time SD-FEA$_{\beta,\gamma,R}$ takes to create a strict improvement. If $m \leq n/2.1$, then*

$$E[T] \leq \binom{n}{m}\left(\frac{1}{1-\gamma} + O\left(\frac{m\ln(R)}{(1-\gamma)n} + R^{-1}\gamma^{-1}\right)\right).$$

Moreover, for all $m \leq n$, we have

$$E[T] = O\left(2^n\frac{\ln(R)}{1-\gamma} + \gamma^{-1}\binom{n}{m}|\lfloor\frac{n}{2.1}\rfloor - m|^{\beta}\right).$$

Theorem 5 provides a good upper bound on the escaping time from a local optimum when there are only few ways to leave it. However, it is not as good when there are many ways to leave the local optimum. The following theorem considers such scenarios. The constant r' defined in the theorem basically represents the first phase that the probability of finding one of the improvements is at least constant, and its value is an integer between 1 and m.

Theorem 6. *Let $\beta > 1$, $0 < \gamma < 1$ and $R \geq e^{1/\gamma}$. Consider the SD-FEA$_{\beta,\gamma,R}$ maximizing a pseudo-Boolean fitness function $f \colon \{0,1\}^n \to \mathbb{R}$. Let $x \in \{0,1\}^n$ be the current search point immediately following a strict improvement or the initial search point. Let $m = \mathrm{FitnessLevelGap}(x)$ and s_m be a lower bound on the probability that a strict improvement is found from search points in the fitness level of x conditional on flipping m bits. Define T as the time SD-FEA$_{\beta,\gamma,R}$ takes to create a strict improvement. If $m \leq n/2.1$, then*

$$E[T] \leq \frac{1}{s_m} \cdot \frac{1}{\gamma}(m - r')^\beta \cdot O\left(1 + \frac{r' \ln(R)}{(1-\gamma)n}\right),$$

where $r' = \min\left\{m, \arg\max_r\left\{\binom{n}{r} \leq \frac{1}{s_m}\frac{1}{\gamma}(m-r)^\beta\right\}\right\}$.

After having established some tools for obtaining upper bounds on the time required to escape from local optima, we now analyze the performance of SD-FEA$_{\beta,\gamma,R}$ on the sub-problems without local optima. On *unimodal functions* the gap of all search points in the search space (except for the global optima) is 1, so the algorithm can always make progress in phase 1.

In the following theorem, we state how SD-FEA$_{\beta,\gamma,R}$ behaves on unimodal functions compared to RLS using an upper bound based on the fitness-level method [22]. The theorem and its proof are similar to the second part of Lemma 4 in [21], and with a good choice of parameters γ and R, the same asymptotic result can be achieved (see the following corollary).

Theorem 7. *Let $\beta > 1$, $0 < \gamma < 1$ and $R \geq e^{1/\gamma}$. Let $f \colon \{0,1\}^n \to \mathbb{R}$ be a unimodal function and $|\mathrm{Im}(f)|$ be the number of its fitness values. Let f_i be the i-th fitness value in an increasing order of the fitness values of f. We consider all fitness levels $A_1, \ldots, A_{|\mathrm{Im}(f)|}$ such that A_i contains search points with fitness value f_i. Let s_i be a lower bound on the probability that RLS finds an improvement from any search point in A_i. Denote by T the runtime of SD-FEA$_{\beta,\gamma,R}$ on f. Then*

$$E[T] \leq \left(\frac{1}{1-\gamma} + O(R^{-1}\gamma^{-1})\right) \sum_{i=1}^{|\mathrm{Im}(f)|-1} \frac{1}{s_i}.$$

The following unimodal benchmark functions ONEMAX and LEADINGONES have been extensively studied in the literature. They are defined by

$$\mathrm{ONEMAX}(x) := \|x\|_1,$$

$$\mathrm{LEADINGONES}(x) := \sum_{i=1}^{n}\prod_{j=1}^{i} x_j$$

for all $x = (x_1, \ldots, x_n) \in \{0,1\}^n$, where $\|x\|_1$ is the number of one-bits in the bit string.

The corollary below is a result of Theorem 7 applied on the unimodal functions ONEMAX with $s_i = (n - (i-1))/n$ and LEADINGONES with $s_i = 1/n$.

Corollary 8. *The expected runtime of the SD-FEA$_{\beta,\gamma,R}$ with $\beta > 1$, $\gamma = o(1)$ and $R \geq e^{1/\gamma}$ on* ONEMAX *is at most* $(1+o(1))n \ln n$ *and on* LEADINGONES *is at most* $(1+o(1))n^2$.

5   Analysis on JUMP$_{k,\delta}$

In this section, we use the results in the previous section to prove a bound on a generalization of JUMP$_\delta$ called JUMP$_{k,\delta}$ with two parameters k and δ, see Fig. 1 for a depiction.

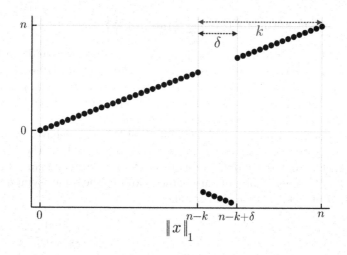

Fig. 1. The function JUMP$_{k,\delta}$.

This function is based on the well-known JUMP benchmark [14], in which the place of the jump with size δ starts at the Hamming distance k from the global optimum. In other words, after the jump, there is a unimodal sub-problem of length $k - \delta$. The classical JUMP function is a special case of JUMP$_{k,\delta}$ with $k = \delta$, i.e., JUMP$_\delta$ = JUMP$_{\delta,\delta}$. Formally, for all $x \in \{0,1\}^n$, we have

$$\text{JUMP}_{k,\delta}(x) = \begin{cases} \|x\|_1 & \text{if } \|x\|_1 \in [0..n-k] \cup [n-k+\delta..n], \\ -\|x\|_1 & \text{otherwise.} \end{cases}$$

We refer the interested reader to see [5] for more information about JUMP$_{k,\delta}$, where the performance of the $(1+1)$ EA, the $(1+1)$ FEA$_\beta$, and the robust version of SD-RLS (SD-RLS^r) are carefully analyzed. Also, Rajabi and Witt [20] independently define the jump function with an offset to analyze the recovery time for the strength in the algorithm SD-RLS with radius memory (SD-RLS^m) after leaving the local optimum. Recently, Witt in [23] analyzes the performance of some other algorithms on the function JUMP$_{k,\delta}$ (which is called JUMPOFFSET in the paper).

We want to show that the algorithm SD-FEA$_{\beta,\gamma,R}$ performs relatively efficiently on JUMP$_{k,\delta}$ in both cases when $k = \delta$ (i. e., JUMP$_\delta$) and $k > \delta$. In the first case, when there is only one improving solution, SD-FEA$_{\beta,\gamma,R}$ with $\gamma = o(1)$ optimizes JUMP$_\delta$ as efficient as SD-RLS^{r} thanks to Theorem 5. The result is formally proven in Theorem 9.

Theorem 9. *The expected runtime $E[T]$ of SD-FEA$_{\beta,\gamma,R}$ with $\beta > 1$, $\gamma = o(1)$ and $R \geq e^{1/\gamma}$ on* JUMP$_\delta$ *with $2 \leq \delta = o(n/\ln(R))$ satisfies*

$$E[T] \leq \binom{n}{\delta}(1 + o(1)).$$

For $\gamma = \Theta(1)$, by closely following the analysis of Theorem 9, it is easy to see that the expected runtime of SD-FEA$_{\beta,\gamma,R}$ on JUMP$_\delta$ is

$$\binom{n}{\delta}\left(\frac{1}{1-\gamma} + o(1)\right),$$

which is still very efficient.

We now present an upper bound on the runtime of the proposed algorithm on JUMP$_{k,\delta}$.

Theorem 10. *The expected runtime $E[T]$ of SD-FEA$_{\beta,\gamma,R}$ with $\beta > 1$, $0 < \gamma < 1$ and $R \geq e^{1/\gamma}$ on* JUMP$_{k,\delta}$ *with $\delta = o(n/\ln(R))$ satisfies*

$$E[T] = O\left(\binom{n}{\delta}\binom{k}{\delta}^{-1}(\delta - r')^\beta \cdot \gamma^{-1} + n\ln n\right),$$

where $r' = \min\left\{\delta, \arg\max_r\left\{\binom{n}{r} \leq \binom{n}{\delta}\binom{k}{\delta}^{-1}\frac{1}{\gamma}(\delta - r)^\beta\right\}\right\}$.

In the following corollary, we see a scenario where we have $r' \geq \delta - c$ for some constant c, resulting in that the term $(\delta - r')^\beta$ disappears from the asymptotic upper bound. This is also an example where the SD-FEA$_{\beta,\gamma,R}$ can asymptotically outperform the (1+1) FEA$_\beta$.

Corollary 11. *Let $\Delta \geq 2$ be a constant. The expected runtime $E[T]$ of SD-FEA$_{\beta,\gamma,R}$ with $\beta > 1$, $0 < \gamma < 1$ and $R \geq e^{1/\gamma}$ on* JUMP$_{k,\delta}$ *with $k = \omega(1) \cap O(\ln n)$ and $\delta = k - \Delta$ satisfies*

$$E[T] = O\left(\binom{n}{\delta}\binom{k}{\delta}^{-1}\gamma^{-1}\right).$$

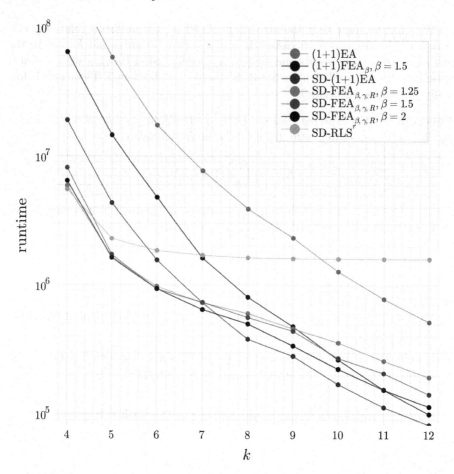

Fig. 2. Average number (over 200 runs) of fitness calls the mentioned algorithms spent to optimize $\text{JUMP}_{k,4}$ ($n = 100$) with different values for k.

6   Experiments

In this section, we present the results of the experiments carried out to measure the performance of the proposed algorithm and several related ones on concrete problem sizes.

We ran an implementation of SD-FEA$_{\beta,\gamma,R}$ with $\beta \in \{1.25, 1.5, 2\}$, $\gamma = 1/4$ and $R = 25$ on the fitness function $\text{JUMP}_{k,\delta}$ of size $n = 100$ with the jump size $\delta = 4$ and k varying from 4 to 13. We recall that we have the classical JUMP function for $k = 4$. We compared our algorithm with the classical $(1 + 1)$ EA with standard mutation rate $1/n$, the $(1+1)$ FEA$_\beta$ from [11] with $\beta = 1.5$, the SD-$(1+1)$ EA presented in [19] with $R = n^2$, and SD-RLS$^\text{r}$ from [21] with $R = n^2$. The parameter settings for these algorithms were all recommended in the corresponding papers. The parameter values for our algorithm were chosen in an ad-hoc fashion, slightly inspired by our theoretical results. All data presented is the average number of fitness calls over 200 runs.

As can be seen in Fig. 2, SD-RLS^r outperforms the rest of the algorithms for $k = 4$, i. e., when there is only one improving solution for local optima. Our SD-FEA$_{\beta,\gamma,R}$ needs roughly $(1 - \gamma)^{-1}$ times more fitness function calls than that since it "wastes" a fraction of γ of the iterations on wrong mutation strengths in phase 4. Not all these iterations are wasted as the small differences for different values of β show. The higher β is, the smaller values the power-law distribution typically takes, meaning that the mutation rate in these iterations stays closer to the ideal one. All three variants of the SD-FEA$_{\beta,\gamma,R}$ significantly outperform the (1+1) FEA$_\beta$, SD-(1+1) EA and $(1 + 1)$ EA. As k is increasing, the average running time of SD-RLS^r improves only little and remains almost without change after $k = 5$; consequently, this algorithm becomes less and less competitive for growing k. This is natural since this algorithm necessarily has to reach phase 4 to be able to flip 4 bits. All other algorithms, especially the (1+1) FEA$_\beta$, perform increasingly better with larger k.

In a middle regime of $k \in \{5, 6, 7\}$, the SD-FEA$_{\beta,\gamma,R}$ has the best average running time among the algorithms regarded. Although both with $k = 4$ and for $k \geq 8$, the SD-FEA$_{\beta,\gamma,R}$ is not the absolutely best algorithm, but its performance loss over the most efficient algorithm (SD-RLS^r for $k = 4$ and SD-(1+1) EA for $k \geq 8$) is small. This finding supports our claim that our algorithm is a good approach to leaving local optima of various kinds.

For a large k, such as 10 or 11, the good performance of the SD-(1+1) EA and (1+1) FEA$_\beta$ might appear surprising. The reason for the slightly weaker performance of our algorithm is the relatively small width of the valley of low fitness ($\delta = 4$), where our algorithm cannot fully show its advantages, but pays the price of sampling from the right heavy-tailed distribution only with probability $\gamma/2$.

7   Recommended Parameters

In this section, we use our theoretical and experimental results to derive some recommendations for choosing the parameters β, γ, and R of our algorithm. We note that having three parameters for a simple $(1 + 1)$-type optimizer might look frightening at first, but a closer look reveals that setting these parameters is actually not too critical.

For the power-law exponent β, as in [11], there is little indication that the precise value is important. The value $\beta = 1.5$ suggested in [11] gives good results even though in our experiments, $\beta = 2$ gave slightly better results. We do not have an explanation for this, but in the light of the small differences we do not think that a bigger effort to optimize β is justified.

Different from the previous approaches building on stagnation detection, our algorithm also does not need specific values for the parameter R, which governs the maximum phase length $\ell_r = \frac{1}{1-\gamma}\binom{n}{r}\ln(R)$ and in particular leads to the property that a single improving solution in distance m is found in phase m with probability $1 - \frac{1}{R}$ (as follows from the proof of Lemma 2). Since we have the heavy-tailed mutations available, it is less critical if an improvement in distance m is missed in phase m. At the same time, since our heavy-tailed mutations also allow to flip more than r bits in phase r, longer phases obtained by taking a larger value of R usually do not have a negative effect on the runtime. For these reasons, the times computed in Theorem 5 depend very little on R. Since the phase length depends only logarithmically on R, we feel that it is safe to choose R as some mildly large constant, say $R = 25$.

The most interesting choice is the value for γ, which sets the balance between the SD-RLS mode of the algorithm and the heavy-tailed mutations. A large rate $1 - \gamma$ of SD-RLS iterations is good to find a single improvement, but can lead to drastic performance losses when there are more improving solutions. Such trade-offs are often to be made in evolutionary computation. For example, the simple RLS heuristic using only 1-bit flips is very efficient on unimodal problems (e.g., has a runtime of $(1 + o(1))n \ln n$ on ONEMAX), but fails on multimodal problems. In contrast, the $(1 + 1)$ EA flips a single bit only with probability approximately $\frac{1}{e}$, and thus optimizes ONEMAX only in time $(1 + o(1))en \ln n$, but can deal with local optima. In a similar vein, a larger value for γ in our algorithm gives some robustness to situations where in phase r other mutations than r-bit flips are profitable – at the price of a slowdown on problems like classic jump functions, where a single improving solution has to be found. It has to be left to the algorithm user to set this trade-off suitably. Taking the example of RLS and the $(1 + 1)$ EA as example, we would generally recommend a constant factor performance loss to buy robustness, that is, a constant value of γ like, e.g., $\gamma = 0.25$.

8   Conclusion

In this work, we proposed a way to combine stagnation detection with heavy-tailed mutation. Our theoretical and experimental results indicate that our new algorithm inherits the good properties of the previous stagnation detection approaches, but is superior in the following respects.

- The additional use of heavy-tailed mutation greatly speeds up leaving a local optimum if there is more than one improving solution in a certain distance m. This is because to leave the local optimum, it is not necessary anymore to complete phase $m - 1$.
- Compared to the robust SD-RLS, which is the fairest point of comparison, our algorithm is significantly simpler, as it avoids the two nested loops (implemented via the parameters r and s in [21]) that organize the reversion to smaller rates. Compared to the SD-$(1 + 1)$ EA, our approach can obtain the better runtimes of the SD-RLS approaches in the case that few improving solutions are available, and compared to the simple SD-RLS of [21], our approach surely converges.

– Again comparing our approach to the robust SD-RLS, our approach gives runtimes with exponential tails. Let m be constant. If the robust SD-RLS misses an improvement in distance m in the m-th phase and thus in time $O(n^m)$ – which happens with probability $n^{-\Theta(1)}$ for typical parameter settings –, then strength m is used again only after the $(m+1)$-st phase, that is, after $\Omega(n^{m+1})$ iterations. If our algorithm misses such an improvement in phase m, then in each of the subsequent $\ell_{m+1} = \Omega(n^{m+1})$ iterations, it still has a chance of $\Omega(n^{-m}\gamma)$ to find this particular improvement. Hence the probability that finding this improvement takes $\Omega(n^{m+1})$ time, is only $(1 - \Omega(n^{-m}\gamma))^{\Omega(n^{m+1})} \leq \exp(-\Omega(n\gamma))$.

As discussed in Sect. 7, the three parameters of our approach are not too critical to set. For these reasons, we believe that our combination of stagnation detection and heavy-tailed mutation is a very promising approach.

As the previous works on stagnation detection, we have only analyzed stagnation detection in the context of a simple hillclimber. This has the advantage that it is clear that the effects revealed in our analysis are truly caused by our stagnation detection approach. Given that there is now quite some work studying stagnation detection in isolation, for future work it would be interesting to see how well stagnation detection (ideally in the combination with heavy-tailed mutation as proposed in this work) can be integrated into more complex evolutionary algorithms.

Acknowledgement. This work was supported by a public grant as part of the Investissements d'avenir project, reference ANR-11-LABX-0056-LMH, LabEx LMH and a research grant by the Danish Council for Independent Research (DFF-FNU 8021-00260B) as well as a travel grant from the Otto Mønsted foundation.

References

1. Antipov, D., Buzdalov, M., Doerr, B.: Fast mutation in crossover-based algorithms. In: Genetic and Evolutionary Computation Conference, GECCO 2020, pp. 1268–1276. ACM (2020)
2. Antipov, D., Buzdalov, M., Doerr, B.: First steps towards a runtime analysis when starting with a good solution. In: Bäck, T., et al. (eds.) PPSN 2020. LNCS, vol. 12270, pp. 560–573. Springer, Cham (2020). https://doi.org/10.1007/978-3-030-58115-2_39
3. Antipov, D., Buzdalov, M., Doerr, B.: Lazy parameter tuning and control: choosing all parameters randomly from a power-law distribution. In: Genetic and Evolutionary Computation Conference, GECCO 2021, pp. 1115–1123. ACM (2021)
4. Antipov, D., Doerr, B.: Runtime analysis of a heavy-tailed $(1 + (\lambda, \lambda))$ genetic algorithm on jump functions. In: Bäck, T., et al. (eds.) PPSN 2020. LNCS, vol. 12270, pp. 545–559. Springer, Cham (2020). https://doi.org/10.1007/978-3-030-58115-2_38
5. Bambury, H., Bultel, A., Doerr, B.: Generalized jump functions. In: Genetic and Evolutionary Computation Conference, GECCO 2021, pp. 1124–1132. ACM (2021)

6. Corus, D., Oliveto, P.S., Yazdani, D.: Automatic adaptation of hypermutation rates for multimodal optimisation. In: Foundations of Genetic Algorithms, FOGA 2021, pp. 4:1–4:12. ACM (2021)

7. Corus, D., Oliveto, P.S., Yazdani, D.: Fast immune system-inspired hypermutation operators for combinatorial optimization. IEEE Trans. Evol. Comput. **25**, 956–970 (2021)

8. Dang, D., Friedrich, T., Kötzing, T., Krejca, M.S., Lehre, P.K., Oliveto, P.S., Sudholt, D., Sutton, A.M.: Escaping local optima using crossover with emergent diversity. IEEE Trans. Evol. Comput. **22**, 484–497 (2018)

9. Doerr, B.: Does comma selection help to cope with local optima? In: Genetic and Evolutionary Computation Conference, GECCO 2020, pp. 1304–1313. ACM (2020)

10. Doerr, B., Doerr, C.: Theory of parameter control for discrete black-box optimization: provable performance gains through dynamic parameter choices. In: Theory of Evolutionary Computation. NCS, pp. 271–321. Springer, Cham (2020). https://doi.org/10.1007/978-3-030-29414-4_6

11. Doerr, B., Le, H.P., Makhmara, R., Nguyen, T.D.: Fast genetic algorithms. In: Genetic and Evolutionary Computation Conference, GECCO 2017, pp. 777–784. ACM (2017)

12. Doerr, B., Rajabi, A.: Stagnation detection meets fast mutation. CoRR abs/2201.12158 (2022). https://arxiv.org/abs/2201.12158

13. Doerr, B., Zheng, W.: Theoretical analyses of multi-objective evolutionary algorithms on multi-modal objectives. In: Conference on Artificial Intelligence, AAAI 2021, pp. 12293–12301. AAAI Press (2021)

14. Droste, S., Jansen, T., Wegener, I.: On the analysis of the $(1+1)$ evolutionary algorithm. Theoret. Comput. Sci. **276**, 51–81 (2002)

15. Friedrich, T., Göbel, A., Quinzan, F., Wagner, M.: Evolutionary algorithms and submodular functions: Benefits of heavy-tailed mutations. CoRR abs/1805.10902 (2018)

16. Friedrich, T., Göbel, A., Quinzan, F., Wagner, M.: Heavy-tailed mutation operators in single-objective combinatorial optimization. In: Auger, A., Fonseca, C.M., Lourenço, N., Machado, P., Paquete, L., Whitley, D. (eds.) PPSN 2018. LNCS, vol. 11101, pp. 134–145. Springer, Cham (2018). https://doi.org/10.1007/978-3-319-99253-2_11

17. Friedrich, T., Quinzan, F., Wagner, M.: Escaping large deceptive basins of attraction with heavy-tailed mutation operators. In: Genetic and Evolutionary Computation Conference, GECCO 2018, pp. 293–300. ACM (2018)

18. Prügel-Bennett, A.: When a genetic algorithm outperforms hill-climbing. Theoret. Comput. Sci. **320**, 135–153 (2004)

19. Rajabi, A., Witt, C.: Self-adjusting evolutionary algorithms for multimodal optimization. In: Genetic and Evolutionary Computation Conference, GECCO 2020, pp. 1314–1322. ACM (2020)

20. Rajabi, A., Witt, C.: Stagnation detection in highly multimodal fitness landscapes. In: Genetic and Evolutionary Computation Conference, GECCO 2021. pp. 1178–1186. ACM (2021)

21. Rajabi, A., Witt, C.: Stagnation detection with randomized local search. In: Zarges, C., Verel, S. (eds.) EvoCOP 2021. LNCS, vol. 12692, pp. 152–168. Springer, Cham (2021). https://doi.org/10.1007/978-3-030-72904-2_10

22. Wegener, I.: Theoretical aspects of evolutionary algorithms. In: Orejas, F., Spirakis, P.G., van Leeuwen, J. (eds.) ICALP 2001. LNCS, vol. 2076, pp. 64–78. Springer, Heidelberg (2001). https://doi.org/10.1007/3-540-48224-5_6

23. Witt, C.: On crossing fitness valleys with majority-vote crossover and estimation-of-distribution algorithms. In: Foundations of Genetic Algorithms, FOGA 2021, pp. 2:1–2:15. ACM (2021)
24. Wu, M., Qian, C., Tang, K.: Dynamic mutation based pareto optimization for subset selection. In: Huang, D.-S., Gromiha, M.M., Han, K., Hussain, A. (eds.) ICIC 2018. LNCS (LNAI), vol. 10956, pp. 25–35. Springer, Cham (2018). https://doi.org/10.1007/978-3-319-95957-3_4

Author Index

Printed in the United States
by Baker & Taylor Publisher Services